MW00724119

Pure Red Cell Aplasia

The Johns Hopkins Series in Contemporary Medicine and Public Health

ALSO OF INTEREST IN THIS SERIES:

Automated Blood Counts and Differentials, J. DAVID BESSMAN, M.D.

Hypoxia, Polycythemia, and Chronic Mountain Sickness, ROBERT M. WINSLOW, M.D., AND CARLOS MONGE C., M.D.

Blood, Blood Products, and AIDS, EDITED BY R. MADHOK, M.D., C. D. FORBES, M.D., AND B. L. EVATT, M.D.

Pure Red Cell Aplasia

Emmanuel N. Dessypris, M.D.

Associate Professor of Medicine, Vanderbilt University, and Staff Hematologist-Oncologist, Veterans Administration Medical Center, Nashville, Tennessee

The Johns Hopkins University Press Baltimore and London

Notice: Readers should scrutinize product information sheets for dosage changes or contraindications, particularly for new or infrequently used drugs.

The Johns Hopkins University Press
701 West 40th Street, Baltimore, Maryland 21211
The Johns Hopkins Press Ltd., London

The paper used in this publication meets the minimum requirements of American National Standard for Information Sciences—Permanence of Paper for Printed Library Materials, ANSI Z39.48-1984.

Library of Congress Cataloging-in-Publication Data

Dessypris, Emmanuel N., 1946–
Pure red cell aplasia / Emmanuel N. Dessypris.
p. cm. — (The Johns Hopkins series in contemporary medicine and public health)
Bibliography: p.
Includes index.
ISBN 0-8018-3572-0 (alk. paper)
1. Pure red cell aplasia. I. Title. II. Series.
[DNLM: 1. Red Cell Aplasia, Pure. WH 155 D475p]
RC641.7.P87D47 1988
616.1′51—dc19
DNLM/DLC
for Library of Congress 87-27378
CIP

Στούς γονείς μου
and to Mary, Margy, and Nicholai

Contents

Figures

Tables

Foreword

Over the past 35 years much information has been acquired on the normal regulation of erythropoiesis. Erythropoietin was completely purified and shown to be the normal regulator of red cell production. Elegant marrow cell tissue culture methods were developed that identified a spectrum of erythroid progenitor cells by their capacity to give rise to different erythroid colonies. A more precise radioimmunoassay for erythropoietin became available and recombinant erythropoietin was produced that was shown to be an effective drug in treating the anemia of renal disease. As this knowledge became available it was natural to ask some new questions about pure red cell aplasia (PRCA). How did individuals suddenly develop a complete inability to make new red cells? Was it a lack of erythropoietin or a failure of the bone marrow? How might this disease be terminated?

Some investigators hypothesized that PRCA could be the result of a lack of erythropoietin, but this was soon dismissed when the bioassay for the hormone showed very high serum levels in most cases. Other investigators infused the plasma or serum of these patients into small animals, but interpretation of the results was hindered by the limited number of patients involved (often only one) and the problem of reaction to foreign proteins. It was the culture of PRCA marrow cells with erythropoietin, and the observation of a marked increase in new hemoglobin and erythroblast formation, that provided new insight into the disease. The remarkable feature of PRCA and transient erythroblastopenia of childhood is that erythroid progenitor cells are present in 60 percent of the patients, and grow normally in vitro, yet will not produce new red cells in the patients' own bodies. This strengthened the idea that an inhibitor of erythropoiesis was present in the patient and, subsequently, in many cases, an IgG inhibitor of autologous erythropoiesis in vitro was identified.

Further research has identified a range of erythroid target cells for the IgG inhibitor and suggested that, in some cases, particularly of chronic lymphocytic leukemia with PRCA, T cells may be involved in the suppression of erythropoiesis. With the advent of information on the pathogenetic mechanism of PRCA, and the knowledge of a variety of concomitant immunologic abnormalities, treatment with a succession of immunosuppressive therapies has been instituted and patients with PRCA now have an 80 percent chance of entering a remission and a 54 percent chance of remaining free of transfusions thereafter. This is likely to improve as new therapies, such as repeated courses of antithymocyte globulin, are given more credence and new combinations of therapy are introduced including cyclosporine.

In this book, Dr. Dessypris has clearly and meticulously recorded the adventures of all who have investigated PRCA. The book is extremely comprehensive and no relevant detail has been omitted. However, this is not simply a profound mapping of information. Close attention has been paid to the relation of the information obtained and in every case sound summary conclusions are drawn so that the reader is brought up to date with current thought on the disease and the reasons for that thought. Areas where present information still does not provide a sound conclusion are clearly exposed, such as the lack of an identifiable antigen for the IgG inhibitor up to the present time and the lack of a precise role for T cells subsets in the pathogenesis of some cases of PRCA.

I have worked closely with Emmanuel Dessypris for ten years, during which time he has contributed much to our new knowledge of PRCA by his fine laboratory and clinical investigation. During that time I have come to appreciate his fertile mind, scholarly devotion to detail, and ability to grasp quickly the essential information from a mass of data. This book is a landmark that will be repeatedly used by all who work in the area, all who take delight in discovery, and all who want to teach others about the disease. The book has a remarkable reference base of the past and will serve as a solid support for the future.

Pure red cell aplasia is a somewhat uncommon disease. Why then should so much time and energy be spent on it? It is very important to the few patients who have the disease. In addition, it raises questions in the mind that human beings do not readily pass off without answers. Most important, it serves as an example for discovery in other diseases. Already a similar pathogenesis has been discovered for pure white cell aplasia and megakaryocyte aplasia. When the site of immunoglobulin attachment to erythroid cells is demonstrated might this not expose a very important element of normal erythropoiesis? Will the disease tell us something special about T cell regulation of erythropoiesis? The precise pathogenesis of aplastic anemia is still not clear, but that disease is responding to immunosuppressive treatment when bone marrow transplantation is not available.

Perhaps knowledge of PRCA will be helpful in finally delineating the pathogenesis of aplastic anemia. We need to comprehend the past in order to discover the future and we must thank Dr. Dessypris for his help with both.

Sanford B. Krantz, M.D.
Vanderbilt University Medical School and
Veterans Administration Medical Center,
Nashville, Tennessee

Acknowledgments

I would like to thank many individuals for their contributions during the various stages of completion of this monograph: Barbara A. Meadows, Sue Miller, Faye Kulp, and the staff of the Veterans Administration Medical Library for their invaluable help in searching the literature; Mary M. Peel and the staff of the Medical Media Production Service for their work on the illustrations; Mary N. Wilson and Mary J. Rich for their secretarial assistance; Mary Lou Kenney for editing the manuscript; Wendy Harris for her helpful advice; Dr. Robert T. Means for his assistance in proofreading; and Dr. Sanford B. Krantz for his continuous interest in this book, his advice, and his support. I would also like to express my appreciation to the staff of the Clinical Research Center at Vanderbilt University for their assistance over the last ten years in the care of patients with pure red cell aplasia.

Pure Red Cell Aplasia

Introduction

Pure red cell aplasia (PRCA) is a syndrome characterized by normocytic normochromic anemia with reticulocytopenia resulting from isolated selective erythroid aplasia in the bone marrow. This syndrome was first described in 1922 by Kaznelson, who reported the case of a 58-year-old man with progressive normochromic anemia characterized by the absence of polychromatophilic erythrocytes and reticulocytes in the peripheral blood and associated with the disappearance of all types of erythroblasts from the bone marrow, but with the maintenance of normal leucopoiesis and thrombopoiesis. He called this disease pure aplastic anemia to differentiate it from aplastic anemia with pancytopenia, which Ehrlich had described 30 years earlier (Kaznelson, 1922). Following its initial description, the same syndrome has been described under different names (table 1), such as erythrophthisis (Baar, 1927; Lupu and Nicolau, 1931), atypical anemia (Kloster, 1934), chronic hypoplastic anemia (Josephs, 1936; Diamond and Blackfan, 1938), aplastic crisis (Owren, 1948), acute erythroblastopenia (Gasser, 1949; Hansen, 1951), erythrogenesis imperfecta (Cathie, 1950), isolated aplastic anemia (Whitby and Britton, 1953), primary or pure red cell aplasia or anemia (Donnelly, 1953; Sakol, 1954; Fontain and Dales, 1955; Gasser, 1955), chronic hypoplastic anemia (Tsai and Levin, 1957), pure red cell agenesis (Schmid et al., 1963), and transient erythroblastopenia of childhood (Wranne, 1970). Today only three of the above terms are in common use: pure red cell aplasia (PRCA) is used to describe this syndrome in adults; congenital hypoplastic anemia (CHA) and transient erythroblastopenia of childhood (TEC) are used to describe the two forms of this syndrome, congenital and acquired, that occur in infants and children.

Pure red cell aplasia in children or adults is an uncommon syndrome the exact incidence of which is not known. It has been suggested, however, that PRCA is more common than the literature indicates, mainly because the majority of cases today are not reported unless they reveal a new association of PRCA with another disease, or an unusual feature of this syn-

Table 1 Synonyms for Pure Red Cell Aplasia

Pure Aplastic Anemia	Isolated Aplastic Anemia
Erythrophthisis	Primary Red Cell Anemia
Atypical Anemia	Refractory Aregenerative Anemia
(Congenital) Hypoplastic Anemia	Transient Erythroblastopenia of Childhood
Aplastic Crisis	Pure Red Cell Agenesis
Acute Erythroblastopenia	Erythroblastopenic Anemia
Erythrogenesis Imperfecta	

drome that has not been previously described. PRCA may affect any age group and occurs with almost the same frequency among males and females. The syndrome has been reported from nearly all parts of the world and there seems to be no specific ethnic or racial predisposition. PRCA may present as a primary hematologic disorder or a secondary hematologic complication of a variety of diseases.

Despite its rarity, PRCA has attracted tremendous interest from many investigators because it represents the most common form of marrow aplasia in which only a single line of hematopoiesis is affected and because of its common association in adults with thymomas and a variety of autoimmune diseases or with a number of laboratory findings indicative of immune dysfunction. The latter has led many investigators to study the possible immune mechanisms through which erythropoiesis is arrested. This has provided increasing evidence for its autoimmune nature and has given the opportunity to study a number of possible mechanisms of immune suppression of erythropoiesis, and hematopoiesis in general, in various human diseases. PRCA is the first human disease in which immune injury to human bone marrow has been documented in a substantial number of patients. In addition, these studies have shown that what morphologically is recognized as an isolated erythroid aplasia pathogenetically is quite heterogeneous regarding the mechanism of immune suppression of erythropoiesis and the target erythroid cell that is predominantly affected. As a reasonable consequence of these studies, immunosuppressive agents have been widely used to treat this disorder, and immunosuppressive therapy has been found to be an effective form of treatment for this syndrome.

In this monograph, I present a detailed description of the wide clinical spectrum of this syndrome, an analysis of the major studies on its pathogenesis, a description of currently available therapeutic approaches and their results, and the natural history of this rare hematologic disorder.

1. Classification

The first classification of pure red cell aplasia (PRCA) was proposed by Gasser in 1957. This classification was based essentially on pediatric cases that had been studied and reported up to that time. Four categories were included: (a) acute erythroblastopenia, including aplastic crises in hemolytic anemias; (b) idiopathic chronic erythroblastopenia; including cases of congenital and acquired PRCA; (c) immune erythroblastopenia; and (d) erythroblastopenia associated with renal failure (Gasser, 1957).

In 1966, DiGiacomo et al. proposed an etiologic classification of PRCA in which they separated the congenital from the acquired forms. They further subclassified the acquired variety into primary cases and cases associated with infections, toxic exposure, neoplastic conditions, hemolytic anemias, uremia, and nutritional deficiencies. This classification was slightly modified and extended by Krantz and Zaentz in 1977, and in its updated form constitutes probably the most clinically useful and currently more frequently used classification of PRCA (table 2).

Congenital Pure Red Cell Aplasia

Congenital pure red cell aplasia (congenital hypoplastic anemia [CHA], Diamond-Blackfan syndrome) was first described by Josephs, Diamond, and Blackfan as a moderate-to-severe chronic anemia beginning in the first year of life, associated with marrow-erythroid hypoplasia but with the maintenance of normal leucopoiesis and thrombopoiesis (Josephs, 1936; Diamond and Blackfan, 1938). This disease is quite rare. More than 200 cases were reported in the literature by 1980, but of those only 93 were well documented (Diamond et al., 1976; Hardisty, 1976). This type of PRCA is thought to be congenital, since 25–57 percent of affected infants are pale at birth, 65–72 percent are anemic by the age of 6 months, and 90 percent are anemic by the first year (Diamond et al., 1976; Alter, 1980). Only in a minority of cases does the disease manifest itself after the first year

Table 2 The Classification of Pure Red Cell Aplasia

- I. Congenital
  - Congenital hypoplastic anemia (Diamond-Blackfan syndrome)
- II. Acquired
  - A. Primary
    1. Autoimmune
    2. Idiopathic
    3. Preleukemic
  - B. Secondary, associated with
    1. Thymoma
    2. Nonthymic solid tumors
    3. Hematologic malignancies
    4. Infections
    5. Chronic hemolytic anemias (aplastic crisis)
    6. Collagen vascular diseases
    7. Drugs and chemicals
    8. Pregnancy
    9. Severe renal failure
    10. Severe nutritional deficiencies
    11. Miscellaneous

Source: Modified from Krantz and Zaentz, 1977.

of life and in exceptional cases after the first 18 months (Hughes, 1961; Diamond et al., 1976; Alter, 1980). Under very unusual circumstances the disease may escape diagnosis early in life and be diagnosed during adulthood (Balaban et al., 1985).

The possible congenital nature of this disorder is supported by the presence of a variety of congenital anomalies in these children (see chapter 2). On the other hand, the hereditary nature is supported by the fact that in at least 15 well-documented cases more than one member of the family was affected; in seven families there were two affected siblings, in another family two affected identical twins, in three other families two affected half-siblings, in three more families the mother and child had the same hematologic condition, and in three additional families the disease was detected in three successive generations (Burgert et al., 1954; Förare, 1963; Seligmann et al., 1963; Mott et al., 1969; Saint-Aimé, 1971; Falter and Robinson, 1972; Hunter and Hakami, 1972; Hamilton et al., 1974; Lawton et al., 1974; Waterkotte and McElfresh, 1974; Diamond et al., 1976; Brody, 1982; Gray, 1982; McFarland et al., 1982. The disease has been reported mainly in Caucasian infants, although cases in black or Asian children have also been described (Ibrahim et al., 1966; Diamond et al., 1976). If indeed this disease is genetically determined, the mode of inheritance has not yet been established (Diamond et al., 1976; Hardisty, 1976).

Primary Acquired Pure Red Cell Aplasia

Primary acquired PRCA is a rare disease that affects individuals of any age without any apparent cause and in the absence of any underlying disorder. Males and females are equally affected and there is no proven racial predisposition (Schmid et al., 1963; DiGiacomo et al., 1966; Krantz and Zaentz, 1977; Krantz and Dessypris, 1985). PRCA in adults is a rare syndrome. By 1953, 28 cases had been reported in the literature (Tsai and Levin, 1957) and 43 by 1963 (Schmid et al., 1963). Between 1972 and 1977, Krantz and Zaentz learned of about 80 new cases that they were able to document by reviewing the bone marrow slides (Krantz and Zaentz, 1977). The exact incidence of the disease is not known.

The primary form of acquired PRCA may run an acute and usually self-limited course or may persist chronically as a form of refractory anemia. In children the same disorder is usually referred to as transient erythroblastopenia of childhood (TEC), an acute form of PRCA of limited duration (Wranne, 1970). In adults, acute idiopathic cases of PRCA are relatively rare, whereas the chronic forms predominate. This is a situation similar to idiopathic thrombocytopenic purpura, the acute self-limited form of which is commonly seen in children and the chronic form in adults. It is conceivable that many cases of acute idiopathic PRCA in adults may pass totally unnoticed, since transient suppression of erythropoiesis causes no symptoms in otherwise normal adults, and no medical attention is sought. In contrast, the disease is more often detected in children, who are commonly seen by pediatricians for relatively minor illnesses during which a moderate anemia may be discovered, the workup of which may lead to the diagnosis of transient erythroblastopenia. A number of cases of both PRCA and TEC have been shown to have an immune pathogenesis and are classified as autoimmune. However, there are a significant number of cases in which all tests available to detect an autoimmune mechanism may be negative. Although the failure to identify an autoimmune mechanism might be related to the low sensitivity of currently available in vitro tests, these cases are more appropriately classified as idiopathic.

Peschle and colleagues proposed a further subclassification of primary acquired PRCA into three types. The first type (type I) is characterized by the presence of high levels of erythropoietin in the patients' plasmas and of an IgG inhibitor of erythropoiesis. Type II PRCA is characterized by low or absent serum erythropoietin and the presence in the serum of an IgG-neutralizing circulating erythropoietin. In type III PRCA, no inhibitors of erythropoiesis or antibodies to erythropoietin are present and the development of acute leukemia seems to be much more frequent; therefore, this type can be considered as a preleukemic myelodysplastic syndrome (Krantz, 1974; Peschle et al., 1975a, 1978). Although this pathogenetic classification of primary acquired PRCA offers some therapeutic direction and a prognosis, neither serum erythropoietin levels nor assays for serum IgG inhibitors of erythropoiesis are available in the common clinical laboratory. In addition, the failure to detect a serum inhibitor cannot automatically classify the case as preleukemic, since a number of such cases have responded to treatment in the same way as cases with a detectable inhibitor (Peschle et al., 1978; Lacombe et al., 1984). Therapeutic decisions cannot rely on the presence or absence of an erythropoietic inhibitor, since the outcome of treatment seems to be the same in patients with detectable or undetectable inhibitors of erythropoiesis in vitro (Lacombe et al., 1984).

Secondary Acquired Pure Red Cell Aplasia

Pure Red Cell Aplasia and Thymoma

The association of PRCA with thymic neoplasms was first described in 1928 by Matras and Priesel. Although many of the initial reports of PRCA associated with thymic tumors failed to document the presence of the erythroblastopenia and/or the thymic origin of the tumor, subsequent case reports have well established the association of PRCA with thymomas (Polayes and Lederer, 1930; Davidshon, 1941; Humphreys and Southworth, 1945; Chediak et al., 1953; Bakker, 1955; Chalmers and Boheimer, 1954; Ross et al., 1954; Thevénard and Marques, 1955; Weinbaum and Thomson, 1955; Hartwell and Mermod, 1957; Lambie et al., 1957; Mielke, 1957; Clarkson and Prockop, 1958; Green, 1958; Jacobs et al., 1959; Kurrein, 1959; Parry et al., 1959; Freeman, 1960; Havard and Scott, 1960; Signier et al., 1960; Bernard et al., 1962*a*; Couespel et al., 1962; Dreyfus et al., 1962; Havard and Parrish, 1962; Jahsman et al., 1962; Andersen and Ladefoged, 1963; Barre et al., 1964; Kough and Barnes, 1964; Radermecker et al., 1964; Roland, 1964; Barnes, 1965; Chapin, 1965; Nicrosini and Taraschi, 1965; Norman, 1965; Schmid et al., 1965; Wendt et al., 1965; Hirst and Robertson, 1967; Baudoin et al., 1968; Vasavada et al., 1973; Houghton and Toghill, 1978; Zeok et al., 1979; Tiber et al., 1981; Socinski et al., 1983). By 1967 there were 66 well-documented cases of PRCA and thymoma reported in the literature (Hirst and Robertson, 1967), and by 1983 their number exceeded 150.

The incidence of PRCA among patients with thymoma has been quite variable among different series, ranging from less than 1 percent to as high as 15 percent (table 3) (Ringertz and Lidholm, 1956; Soutter et al., 1957; Jacobs et al., 1959; Lattes, 1962; Roland, 1964; Rubin et al., 1964; Legg and Brady, 1965; LeGolvan and Abell, 1977). During a 44-year period at Sidney Hospital in Australia there were only 13 cases of thymic tumors, and in only 3 of them was pure red cell aplasia diagnosed, among 92,000 surgical

Table 3 Incidence of Pure Red Cell Aplasia among Hospitalized Patients with Thymoma

Institution	Years	No. of Thymomas	No. of PRCA
Mayo Clinic	−1960	169	4
University of Sidney Hospital	1919–65	18	3
Johns Hopkins Hospital	1931–71	65	4
University of Michigan Hospital	1938–72	46	0
University of Pennsylvania Hospital	1955–63	8	2
Massachusetts General Hospital	1966–78	40	0
Total		346	13

specimens and necropsies (Hirst and Robertson, 1967). At Mayo Clinic, among 169 cases of thymoma seen until 1960, PRCA was found in only 4 (Schmid et al., 1965). More recently, LeGolvan and Abell, reviewing their experience with thymomas at the University of Michigan from 1938 to 1972, reported 46 cases of thymoma but no case of PRCA was noted among them. It seems, therefore, that PRCA is a rather uncommon manifestation of thymomas, probably not seen in more than 5 percent of cases.

The incidence of thymoma among patients presenting with PRCA has been reported to be as high as 50 percent (Jacobs et al., 1959; Andersen and Ladefoged, 1963; DiGiacomo et al., 1966; Hirst and Robertson, 1967). However, in the Vanderbilt University series of 43 consecutive patients with chronic acquired PRCA, only 3 had thymomas; a fourth patient had had a thymoma removed surgically before the development of red cell aplasia (Krantz and Dessypris, 1985, and unpublished observations). It is our impression that the 50 percent incidence of thymoma among patients with PRCA is based on the early literature, written when emphasis on the association of PRCA and thymoma was strong, and may be an overestimate.

Almost all cases of PRCA and thymoma have occurred in adults, mainly females, over 40 (Hirst and Robertson, 1967). The occurrence of thymomas in children and young adults is rare. The tumors found in younger people are much more aggressive than those found in adults and are seldom associated with any one of the paraneoplastic syndromes commonly described in association with these tumors (Batata et al., 1974). By 1979, only 19 well-documented cases of thymoma in patients younger than 18 years of age had been reported (Bowie et al., 1979). Only a single case of thymoma and PRCA in a child has been described (Talerman and Amigo, 1968). In this case PRCA was of short duration and evolved soon after diagnosis into aplastic anemia. The most frequent histologic types of thymoma associated with PRCA are the spindle cell and mixed spindle cell-lymphocytic varieties (Schmid et al., 1965; Hirst and Robertson, 1967), although PRCA has also been described in pure lymphocytic and mixed lymphoepithelial thymomas, as well as in association with thymolipoma (Shillitoe and Goodyear, 1960; Lebrun et al., 1985). No correlation has been firmly established between the histologic type of thymoma and erythroid aplasia or any other paraneoplastic systemic manifestation of these tumors. The malignant or benign nature of thymomas does not correlate with distinct cytologic features of the tumor cells but with the invasive characteristics of the tumor (Rosai and Levine, 1976). In more than 50 percent of the cases, the tumor is well encapsulated and there is no evidence of metastatic disease or invasion of adjacent thoracic structures. In the majority of cases of PRCA associated with thymoma, the tumor is not invasive or metastatic (Hirst and Robertson, 1967). However, PRCA has been reported in association both with "benign" and invasive or metastatic thymoma (Migueres et al., 1968; Rogers et al., 1968; Whitaker et al., 1970; DeSevilla et al., 1975). The thymoma can precede the development of PRCA by many years, or PRCA may develop many years after surgical removal of the tumor (Dameshek et al., 1967; Hirst and Robertson, 1967; Albahary, 1977; Varet et al., 1978).

Pure Red Cell Aplasia and Nonthymic Solid Tumors

A variety of nonhematologic, nonthymic neoplasms have been reported in association with acquired PRCA (Doll and Weiss, 1985) (table 4), including adenocarcinoma of the stomach (Kark, 1937; Gajwani and Zinner, 1976), breast (Clarkson and Prockop, 1958; Green, 1958; Slater et al., 1979), adenocarcinoma of the bile ducts (Lee et al., 1980), bronchogenic carcinoma (Tsai and Levin, 1957; Entwistle et al., 1964; Brafield and Verbov, 1966; Guthrie and Thornton, 1983), infiltrative squamous cell carcinoma of the skin (Guthrie and Thornton, 1983), Kaposi's sarcoma (Hirst and Robertson, 1967; Marmont et al., 1975; Shimm et al., 1979), thyroid carcinoma (Iannuci et al., 1983), and adenocarcinoma of unknown primary site (Mitchell et al., 1971). In a number of these cases a thymoma was also detected on autopsy, so the association of PRCA with the nonthymic neoplasm remains questionable. Considering the rarity of reports of malignant neoplasms associated with PRCA and the high incidence of malignancies, it is evident that the association of PRCA with nonthymic, nonhematologic malignancies is extremely rare, raising questions about the mere existence of such an association. Alternatively, one might argue that PRCA is underdiagnosed in the presence of malignant neoplasms because severe anemia is such a common manifestation of malignant tumors that bone marrow examination is not routinely included in the investigation of anemia in the majority of cases (Gajwani and Zinner, 1976). Whether PRCA is a true paraneoplastic syndrome, as certain authors have suggested (Lee et al., 1980), is not clear, since the activity of the neoplasia does

Table 4 Pure Red Cell Aplasia and Nonthymic Solid Tumors

Malignant Neoplasm	Number of Cases	Authors
Gastric adenocarcinoma	2	Kark, 1937; Gajwani and Zinner, 1976
Breast carcinoma	3	Green, 1958; Clarkson and Prockop, 1958; Slater et al., 1979
Bile duct adenocarcinoma	1	Lee et al., 1980
Bronchogenic carcinoma	4	Tsai and Levin, 1957; Entwistle et al., 1964; Brafield and Verbov, 1966; Guthrie and Thornton, 1983
Squamous cell carcinoma of skin (infiltrative)	1	Guthrie and Thornton, 1983
Kaposi's sarcoma	3	Hirst and Robertson, 1967; Marmont et al., 1975; Shimm et al., 1979
Thyroid carcinoma	1	Iannuci et al., 1983
Unknown primary	1	Mitchell et al., 1971

not correlate with the activity of PRCA, and, with a single exception in all reported cases, treatment of underlying malignancy did not lead to remission of PRCA. In general, PRCA has been reported to appear a few months to three years before the diagnosis of carcinoma and tends to follow a chronic course independent of the evolution of the underlying malignant tumor.

Pure Red Cell Aplasia and Hematologic Malignancies

Various hematologic malignancies have been associated with PRCA (table 5). In many of these conditions, in which there is marrow involvement by neoplastic hematopoietic cells, the term *erythroid aplasia* or *red cell aplasia* seems more appropriate than pure red cell aplasia.

CHRONIC LYMPHOCYTIC LEUKEMIA

Chronic lymphocytic leukemia (CLL) is the most frequently reported hematologic malignancy associated with red cell aplasia (table 5) (Battle et al., 1963; Stohlman et al., 1971; Tatarsky, 1972; Abeloff and Waterbury, 1974; Hoffman et al., 1978; Blanc, 1980; Katz et al., 1981; Mangan et al., 1981, 1982; Nagasawa et al., 1981; Viala et al., 1981; Clauvel et al., 1983; Lipton et al., 1983; Thiagarajan et al., 1983; Yoo et al., 1983; Newland et al., 1984; Lebrun et al., 1985; Soler et al., 1985; Chikkappa et al., 1986*b*; Hansen et al., 1986). It has been estimated that the incidence of PRCA among patients with CLL may be as high as 6 percent (Chikkappa et al., 1986*b*). By 1986, 44 cases of erythroid aplasia and chronic lymphocytic leukemia had been reported, including the common B cell CLL and the rare type of T cell CLL. Twenty-seven of these cases have been reviewed by

Table 5 Erythroid Aplasia and Hematologic Malignancies

Type of Malignancy	Number of Cases	Sources
Chronic lymphocytic leukemia (CLL)		
Type undefined	11	Battle et al., 1963; Stohlman et al., 1971; Tatarsky, 1972; Abeloff and Waterbury, 1974
B cell type	23	Blanc, 1980; Katz et al., 1981; Mangan et al., 1981, 1982; Viala et al., 1981; Clauvel et al., 1983; Thiagarajan et al., 1983; Yoo et al., 1983; Lebrun et al., 1985; Chikkappa et al., 1986
T cell type	10	Hoffman et al., 1978; Callard et al., 1981; Nagasawa et al., 1981; Hocking et al., 1983; Lipton et al., 1983; Newland et al., 1984; Shionoya et al., 1984; Soler et al., 1985; Hansen et al., 1986
Chronic granulocytic leukemia (CGL)	7	Schmid et al., 1963; Bottcher et al., 1970; Kitahara, 1978; Dessypris et al., 1981; Lacombe et al., 1984
Myelofibrosis with myeloid metaplasia	11	Bentley et al., 1977; Clement, 1979; Barosi et al., 1983; Lacombe et al., 1984; Dessypris and Krantz, unpublished observations
Malignant lymphomas		
Hodgkin's lymphoma	4	Bove, 1956; Field et al., 1968; Guerra et al., 1969; Morgan et al., 1978
Non-Hodgkin's lymphomas	8	Jacobs et al., 1959; Jepson and Vas, 1974; Hunt and Lander, 1975; Prasad et al., 1976; Lee et al., 1978; Carloss et al., 1979; Clauvel et al., 1983
Multiple myeloma	1	Gilbert et al., 1958
Acute lymphoblastic leukemia	5	Krantz and Zaentz, 1977; Sallan and Buchanan, 1977; deAlarcon et al., 1978; Imamura, 1986

Chikkappa and associates (1986*b*). It should be noted that in a number of these cases a coexisting thymoma has also been described. The initially reported cases did not define the type of lymphocytes involved in the leukemic process. Most of these early recognized cases were associated with the common B cell CLL. More recently an increasing number of cases of PRCA associated with T cell CLL have been reported. This type of chronic lymphocytic leukemia is frequently associated with cytopenias with selective aplasia of one or more hematopoietic cell lines in the marrow and the presence of an increased number of large granular lymphocytes in the marrow and the peripheral blood (Newland et al., 1984). Recently it has become clear that these cases of T cell lymphocytosis represent monoclonal T cell lymphoproliferative disorders as indicated by rearrangements in the T cell receptor gene (Berliner et al., 1986) and therefore can be considered as cases of T cell CLL. Red cell aplasia was diagnosed in the presence of CLL either at presentation or at any time (months to many years) after the diagnosis of lymphatic leukemia. In a number of patients, recurrent episodes of erythroblastopenia and severe reticulocytopenic anemia have been noted during the course of chronic lymphocytic leukemia. Autoimmune hemolytic anemia with positive antiglobulin reaction and coexisting erythroid aplasia was reported in six cases. Erythroid aplasia has developed in treated or untreated cases of CLL, and discontinuation of chlorambucil or of any other chemotherapeutic agent did not always lead to recovery of erythropoiesis, so treatment with cytotoxic agents cannot always be held responsible for the induction of the erythroid aplasia. However, in five cases, PRCA has been associated with treatment of CLL with chlorambucil, and in a number of these, recovery of erythropoiesis was observed within a short period of time after discontinuation of chlorambucil. A second episode of PRCA was also seen in some cases after reinstitution of chlorambucil treatment (Yoo et al., 1983). It seems that in a small number of cases PRCA in CLL may be induced by chlorambucil; however, in the majority of cases no association with previous cytotoxic therapy can be established. In view of the fact that PRCA in CLL has been successfully treated with chlorambucil (reviewed by Chikkappa et al., 1986*b*), the role of this cytotoxic agent in inducing PRCA in CLL remains unclear. It has been suggested that erythroid aplasia in chronic lymphocytic leukemia may be more frequent than the literature indicates, mainly because severe normochromic anemia with low reticulocyte count is a frequent manifestation of end stage CLL, and in the absence of hemolysis it is usually attributed to the myelophthisic process in the marrow (Viala et al., 1981). A pathogenetic relation may exist between the stage of CLL and the development of erythroid aplasia (Mangan and D'Alessandro, 1985), although development of PRCA does not seem to affect the prognosis of lymphocytic leukemia (Chikkappa et al., 1986*b*). The type of the malignant lymphocytes seems also to play a role in the development of erythroid aplasia, which appears to be relatively more common among cases of T cell CLL (see chapter 4).

CHRONIC GRANULOCYTIC LEUKEMIA

Seven cases of erythroid aplasia associated with chronic granulocytic leukemia (CGL) have thus far been reported (table 5). In one of these, erythroid aplasia preceded the diagnosis of CGL by three years (Schmid et al., 1963), in two cases red cell aplasia was found at the time of diagnosis (Bottcher et al., 1970; Dessypris et al., 1981), and in the remaining four cases erythroblastopenia with severe anemia developed during the course of CGL and after treatment with busulfan (Kitahara, 1978; Dessypris et al., 1981; Lacombe et al., 1984). In the last two cases no change in erythropoiesis was noted after discontinuation of busulfan, so this agent was not considered responsible for the erythroid aplasia. Erythroid aplasia in the course of CGL has been considered to be an early change preceding the terminal phase of blastic transformation (Spiers, 1976); however, the reported cases in the literature do not support this view. The development of red cell aplasia before, at the same time, or after the diagnosis of CGL seems to indicate that this syndrome may have little or no relation to CGL (Dessypris et al., 1981).

MYELOFIBROSIS AND MYELOID METAPLASIA

Anemia is one of the most frequent manifestations of this myeloproliferative disorder (Ward and Block, 1971), and in a small number of cases, not exceeding 6 percent, erythroid hypoplasia of variable severity, mainly identifiable by ferrokinetic studies, has repeatedly been reported (Meeus-Bith et al., 1957; Nathan and Berlin, 1959; Bowdler, 1961; Szur and Smith, 1961; Nakkai et al., 1962; Oeltgen and Pribilla, 1964; Brunner, 1965; Kesse-Elias et al., 1968; Milner et al., 1973). However, complete erythroid aplasia is a rare event and red cell aplasia associated with myelofibrosis has been reported in only nine cases (Bentley et al., 1977; Clement, 1979; Barosi et al., 1983; Lacombe et al., 1984). Two more cases of erythroid aplasia associated with myelofibrosis were also seen at the Vanderbilt University Hospital between 1976 and 1984 (Dessypris and Krantz, unpublished observations). In all reported cases severe anemia with reticulocytopenia, a normal white cell and platelet count, and a normal or mildly increased spleen size were the major hematologic findings. In eight cases the diagnosis of both myelofibrosis and red cell aplasia was made at the first marrow examination, in another case red cell aplasia preceded the development of typical myelofibrosis, and in one of our patients red cell aplasia developed almost three years after the diagnosis of myelofibrosis. Most of the patients with these two disorders were relatively younger than the average patient with myelofibrosis and myeloid metaplasia. In those patients in whom a splenectomy was performed, there was only minimal extramedullary hematopoiesis on histologic examination of the spleen with noticeably absent erythroid component. In two of the reported cases the clinical and hematologic findings were indistinguishable from those described as typical for malignant myelosclerosis (Lewis and Szur, 1963). A poorer survival rate and a tendency toward blastic transformation were seen in the majority of the reported cases. Whether the reported cases represent a coincidence of red cell aplasia and myelofibrosis or whether there is a true association between these two hematologic disorders is not known.

MALIGNANT LYMPHOMAS

PRCA has been reported in association with malignant lymphomas as another autoimmune syndrome seen in the course of malignancies of the lymphopoietic tissue. Four cases of Hodgkin's lymphoma and PRCA have been described (Bove, 1956; Field et al., 1968; Guerra et al., 1969; Morgan et al., 1978). In one of them (Field et al., 1968), a coexisting thymoma was also found; in another case PRCA preceded the development of autoimmune hemolytic anemia (Guerra et al., 1969). Usually remission of PRCA occurred with treatment of Hodgkin's lymphoma. However, such a treatment is highly immunosuppressive, and the remission of PRCA might have been the result of elimination of lymphoma or the result of the immunosuppression per se, so the relation of PRCA with the underlying lymphoma is not known.

Eight cases of non-Hodgkin's lymphomas and concurrent red cell aplasia have also been described (Jacobs et al., 1959; Jepson and Vas, 1974; Hunt and Lander, 1975; Lee et al., 1978; Carloss et al., 1979; Clauvel et al., 1983). The lymphoma was of the poorly differentiated lymphocytic type in six cases and of the histiocytic (large transformed) cell type in two. Again, in all cases the erythroid aplasia remitted with treatment of the malignant lymphoma. In the case reported by Carloss et al. recurrent erythroid aplasia was seen with recurrence of the lymphoma, and there was a relation between marrow infiltration by the lymphoma and erythroid aplasia in the sense that erythroid aplasia was seen only with marrow involvement but not with extramedullary relapse, suggesting an effect of malignant lymphoid cells on erythropoiesis. In another case, however, recurrence of histiocytic lymphoma did not lead to recurrence of PRCA, which had been present at the time of initial diagnosis (Hunt and Lander, 1975). Due to the limited number of reported cases and their short follow-up it is difficult to comment on the nature of the relation of PRCA to various lymphoid malignancies, but, in general, PRCA is viewed as one of the various autoimmune hematologic complications that have been described in malignant lymphoproliferative disorders (Sacks, 1974).

OTHER HEMATOLOGIC MALIGNANCIES

Erythroid aplasia has been reported in a single case of multiple myeloma in which a coexisting thymoma was present (Gilbert et al., 1968) It is more reasonable to consider this case of PRCA as related to thymoma rather than to multiple myeloma.

Selective erythroid aplasia has also been reported in five children with acute lymphoblastic leukemia and in an adult with pre–T cell acute lymphoblastic leukemia. In one of these children and in the case of the adult, PRCA persisted for four months before the diagnosis of acute lymphoblastic leukemia was made and it was considered a preleukemic manifestation (deAlarcon et al., 1978; Imamura et al., 1986). In the remaining four cases, PRCA developed during maintenance chemotherapy for acute lymphoblastic leukemia (ALL); three of the children were in remission and the fourth had recurrent leukemia at the time of diagnosis of PRCA (Sallan and Buchanan, 1977; Lukens JN, personal communication, 1981). Whether these cases represent drug-induced PRCA, are related pathogenetically to acute leukemia, or represent a coincidence of transient erythroblastopenia of childhood in patients on maintenance chemotherapy for ALL is not known.

Pure Red Cell Aplasia and Infections

A number of bacterial and viral infections have been associated with bone marrow depression, resulting in various degrees of pancytopenia. More specifically, suppression of red cell production as indicated by a low reticulocyte count has been described in 20 percent of children during the course of different viral illnesses (Biemer and Taylor, 1982). A transient acute form of erythroid aplasia has been reported in association with meningococcemia, straphylococcemia, atypical pneumonia, mumps, viral hepatitis, infectious mononucleosis, and cytomegalovirus infection (Chernoff and Josephson, 1951; Gasser, 1957; Sears et al., 1974; Wilson et al., 1980; Sao et al., 1982; Purtilo et al., 1984; Socinski et al., 1984; Bruyn and Shelfhout, 1986). It should also be noted that in most cases of transient erythroblastopenia of childhood there is a history of preceding viral illness (Alter and Nathan, 1979). The pathogenetic relation between infections and selective suppression of erythropoiesis is not understood. Recently infection with parvovirus with or without the classical manifestations of erythema infectiosum has been linked with the aplastic crisis seen in patients with various forms of chronic hemolytic anemias (*vide infra*).

Aplastic Crisis in Chronic Hemolytic Anemias

Aplastic crisis constitutes an acute and transient form of PRCA that has been described in almost all forms of congenital and acquired chronic hemolytic anemias including sickle cell anemia (Singer et al., 1950; Chernoff and Josephson, 1951; Leiken, 1957; MacIver and Parker-Williams, 1961; Charney and Miller, 1964), congenital spherocytosis (Dameshek and Bloom, 1948; Owren, 1948; Chanarin et al., 1962; Bouroncle, 1969), pyruvate kinase deficiency (Nixon and Buchanan, 1967; Duncan et al., 1983), paroxysmal nocturnal hemoglobinuria (Crosby, 1953; Pavlic and Bouroncle, 1965), and immune hemolytic anemia (Davis et al., 1952; Eisemann and Dameshek, 1954; Meyer and Bertcher, 1960; Pirofsky, 1969; Meyer et al., 1978). These conditions have in common a shortened red cell survival, so even minor changes in the rate of red cell production can result in rapid and significant changes in the circulating red cell mass. Acute exacerbation of the chronic anemia in a chronic hemolytic disorder, when associated with reticulocytopenia, should lead the physician to a working diagnosis of aplastic crisis. This diagnosis must be confirmed by bone marrow examination, which demonstrates a total absence of erythroblasts, a finding strikingly different from the various degrees of erythroid hyperplasia commonly seen in chronic hemolytic disorders. The combination of hemolysis and reticulocytopenia is not by itself sufficient for the diagnosis of aplastic crisis (Pirofsky, 1969). The crises are self-limited, usually lasting five to ten days. The frequency of aplastic crisis among the different types of hemolytic anemias is not known. In a single study it has been estimated that 10 percent of admissions to the hospital for sickle cell disease are for an aplastic crisis (Charney and Miller, 1964).

It has long been suspected that these cases of acute transient erythroblastopenia have an infectious etiology. Episodes of erythroblastopenia are clustered in time and within families (Dameshek, 1941; Dameshek and Bloom, 1948; Leiken, 1957; Bouroncle, 1969), and very frequently affected individuals give a history of symptoms of "viral illness" preceding the episode of erythroid aplasia. In addition, aplastic crises are almost limited to children, being uncommon in patients older than 15 with hereditary hemolytic anemias (Serjeant et al., 1981). This fact, along with the observation that aplastic crisis seldom occurs for a second time in the same patient, points toward a specific infectious agent responsible for the induction of erythroid aplasia. Recently it has been demonstrated that in the majority of cases of sickle cell disease, hereditary spherocytosis, and a limited number of cases of pyruvate kinase deficiency a human parvovirus causing the fifth disease (exanthema infectiosum) is also responsible for the attacks of transient erythroid aplasia (Pattison et al., 1981; Serjeant et al., 1981; Anderson et al., 1982; Duncan et al., 1983; Kelleher et al., 1983; Rao et al., 1983; Bone marrow, 1983).

Collagen Vascular Diseases

SYSTEMIC LUPUS ERYTHEMATOSUS AND ERYTHROID APLASIA

Pure red aplasia is a rare hematologic complication of systemic lupus erythematosus (SLE), although it is not widely recognized as such; in most reviews on the hematologic aspects of SLE one finds that PRCA is not discussed (Harvey et al., 1954; Budman and Steinberg, 1977; Schreiber, 1980). Erythroid hypoplasia (less than 10 percent erythroid cells) was first noted in 3 out of 38 bone marrow aspirates from patients with SLE (Michael et al., 1951). Complete erythroid aplasia has thus far been reported in at least 11 cases in association with SLE (Kough and Barnes, 1964; Daughaday, 1968; Cassileth and Myers, 1973; Chollet et al., 1973; MacKechnie et al., 1973; Takigawa and Hayakawa, 1974; Cavalcant et al., 1978; Meyer et al., 1978; Oren and Cohen, 1978; Isbister et al., 1981; Francis, 1982; Heck et al., 1985). In two cases PRCA was the only presenting manifestation of SLE, whereas in the remaining cases it presented during the course of already diagnosed SLE. In two cases there was a coexistence of thymoma, SLE, and PRCA in the same patient, and in a third case PRCA developed in repeated episodes in a patient with SLE and autoimmune hemolytic anemia. In the remaining seven patients no thymoma could be detected to be held responsible for the erythroid aplasia. It should be noted that patients with primary PRCA as well as PRCA associated with thymoma frequently have serologic manifestations suggestive of SLE, such as high titer of ANA or positive LE phenomenon, but they also frequently lack the clinical manifestations of SLE (Funkhouser, 1961; Jahsman et al., 1962; Holborow et al., 1963; Kough and Barnes, 1964; Barnes, 1965; Hirst and Robertson, 1967; Hamilton and Conley, 1969; Safdar et al., 1970; Zoupanos, 1970; Socinski et al., 1983). Patients with such abnormalities in the course of PRCA who also have positive antibodies to double-stranded DNA will have to be considered as having SLE presenting with PRCA as its first manifestations (Heck et al., 1985). It may be that in a number of these cases PRCA represents another autoimmune hematologic manifestation associated with infraclinical SLE (Favre et al., 1979).

RHEUMATOID ARTHRITIS

The first patient with rheumatoid arthritis and PRCA was described by Dameshek et al. (1967). Two additional cases of chronic PRCA that developed in patients with long-standing rheumatoid arthritis have been reported (Chang et al., 1978; Konwalinka et al., 1983), but in these reports the authors did not associate the erythroid aplasia with the rheumatoid arthritis. In three other cases of rheumatoid arthritis the syndrome of PRCA appeared after treatment with either gold, fenoprofen, or phenylbutazone and remitted shortly after discontinuation of these drugs; therefore, they were considered as cases of drug-induced PRCA (Swineford et al., 1958; Reid and Patterson, 1977; Weinberger, 1979). Pure red cell aplasia in association with rheumatoid arthritis has been reported in one child and three adults (Rubin et al., 1978; Dessypris et al., 1984). In these cases PRCA ran a chronic course and did not remit following discontinuation of treatment for rheumatoid arthritis. The feature shared by all adult patients with PRCA complicating rheumatoid arthritis, regardless of whether the authors associated the two diseases, was the presence of arthritis of long duration—20 or more years. It seems that drug-induced PRCA may appear at any time during the course of rheumatoid arthritis, but PRCA as a complication of rheumatoid arthritis itself behaves like other extra-articular manifestations of this disease in that it occurs in patients with rheumatoid arthritis of long duration (Dessypris et al., 1984). This is in contrast to the much more commonly seen anemia of chronic disease, the severity of which correlates with the erythrocyte sedimentation rate and the disease activity level but not with the duration of the rheumatoid arthritis (Engstead and Strauberg, 1966).

Drug-induced Pure Red Cell Aplasia

A continuously increasing number of drugs and chemicals have been implicated as causes of PRCA (table 6). In general, drug-induced PRCA is an acute, transient form of erythroblastopenia that remits soon after discontinuation of the responsible drug or following cessation of exposure to the chemical agent. As with other drug-induced allergic reactions, PRCA may appear a short time after initiation of therapy or after years of exposure to the drug or chemical. In most cases the association of PRCA with various drugs is circumstantial and is based on single case reports. In other cases, such as with azathioprine, chloramphenicol, chloropropamide, diphenylhydantoin, and isoniazid there are more than single case reports that make such an association very likely. Diphenylhydantoin, in particular, has been well proven to cause pure red cell aplasia. In the case reported by Brittingham et al. (1964) and Yunis et al. (1967) rechallenge of the patient with diphenylhydantoin on multiple occasions led to the development of erythroid hypoplasia or aplasia that remitted after discontinuation of this agent. Azathioprine also has been shown to cause recurrent erythroid aplasia in patients rechallenged with very small doses (McGrath et al., 1975).

The suspicion that a particular drug or chemical agent is the cause of PRCA is increased when withdrawal of a single medication or cessation of the exposure to a single chemical agent is followed by rapid return of erythropoiesis to normal. However, as has been pointed out by Geary (1978), it is difficult in individual cases to establish a cause and effect relationship between a single drug and this hematologic complication: patients usually take more than one drug, all of which are discontinued before the diagnosis of PRCA; they occasionally suffer from diseases that are by themselves

Table 6 Drugs and Chemicals Associated with Pure Red Cell Aplasia

Drugs and Chemicals	Sources
Allopurinol	Vohra et al., 1983
Aminopyrine	Gasser, 1957
Arsphenamine	Sharff and Neumann, 1944
Azathioprine	McGrath et al., 1975; Old et al., 1978; DeClerck et al., 1980; Flury and Montandon, 1980
Benzene hexachloride	Vodopick, 1975
Bromsulphalein (BSP)	Broccia and Dessalvi, 1983
Calomel	Gasser, 1957
Carbamazepine	Hirai, 1977
Cephalothin	MacCulloch et al., 1974
Chenopodium	Gasser, 1957
Chloramphenicol	Ozer et al., 1960; Hirst and Robertson, 1967; Vilan et al., 1973
Chloropropamide	Recker and Hynes, 1969; Teoh et al., 1973; Gill et al., 1980; Planas et al., 1980
Co-trimoxazole	Stephens, 1974
Diphenylhydantoin	Brittingham et al., 1964; Jeong et al., 1974; Huijgens et al., 1978; Lee et al., 1978; Hotta et al., 1980; Sugimoto et al., 1982; Dessypris et al., 1985
Fenoprofen	Weinberger, 1979; Reitz and Bottomley, 1984
Gold	Reid and Patterson, 1977
Halothane	Jurgensen et al., 1970
Isoniazid	Mielke, 1958; Goodman and Block, 1964; Hoffman et al., 1983; Claiborne and Dutt, 1985
Maloprim (Dapsone and Pyrimethamine)	Nicholls and Concannon, 1982
Mepacrine	Foy and Kondi, 1953*a*
Methazolamide	Krivoy et al., 1981
Oestrogens	Geary, 1978
Penicillin	Gasser, 1949
d-Penicillamine	Gollan et al., 1976
Pentachlorophenol	Roberts, 1983
Phenobarbital	Gasser, 1949
Phenylbutazone	Swineford et al., 1958; Ibrahim et al., 1966
Procainamide	Giannone et al., 1987
Salicylazosulfapyridine	Peschle et al., 1978
Santonin	Gasser, 1957
Sodium dipropylacetate	Douchain et al., 1980
Sodium valproate	MacDougall-Lorna, 1982
Sulfasalazine	Dunn and Kerr, 1981
Sulfathiazol	Strauss, 1943; Tsai and Levin, 1957
Sulindac	Sanz et al., 1980
Thiamphenicol	Cornet et al., 1974; Estavoyer et al., 1981
Tolbutamide	Schmid et al., 1963

associated with this syndrome; and, finally, PRCA has a tendency to remit spontaneously, especially when it appears in its acute form. Therefore, in many single case reports it is difficult to prove whether the PRCA was indeed drug-induced, associated with the patient's underlying disease, or a representation of an acute and transient form of this syndrome.

In a case of suspected drug-induced pure red cell aplasia after discontinuation of the responsible medication erythropoiesis must be expected to return to normal within a period of a few weeks. Theoretically the length of this period will depend on the half-life of the responsible drug and the pathogenetic mechanism responsible for the PRCA. If after three or four weeks of withdrawal of the responsible agent there is no evidence of recovering erythropoiesis as indicated by reappearance of reticulocytes in the peripheral blood and/or maintenance of a stable hematocrit, other causes of PRCA should be considered.

Nutritional Deficiencies and Erythroblastopenia

Erythroid aplasia has been reported in children and adults with severe multiple nutritional deficiencies and kwashiorkor (Kho, 1957; Foy et al., 1961; Kho et al., 1962; Walt et al., 1962; Kondi et al., 1963; Neame and Simpson, 1964; Zucker et al., 1971). In most reported cases the patients also had concurrent infections of various etiologies. Erythroid aplasia was diagnosed either on admission to the hospital or a few weeks later during the recovery period from the malnutrition and infections. This form of PRCA in the reported cases was characterized by the presence of a small number of giant vacuolated proerythroblasts and ran a variable course lasting from two days to more than two months (Kho, 1957; Kho et al., 1962). Treatment of underlying infection and correction of hypoproteinemia and other nutritional deficiencies resulted in correction of the erythroblastopenia diagnosed on admission. In 10 to 20 percent of the cases erythroblastopenia developed during the second or third week of hospitalization after the administration of a high-protein diet, indicating that protein deficiency alone cannot be implicated as the cause of isolated erythroid aplasia (Kho et al., 1962; Zucker et al., 1971). Supplementation of the high-protein diet with folic acid did not prevent the late development of erythroid aplasia (Zucker et al., 1971). The nutritional factor(s) responsible for the erythroid aplasia in these cases of kwashiorkor is not known. Based on experimental data and clinical observations riboflavin deficiency has been considered an essential factor in cases of severe malnutrition associated with erythroid hypoplasia and a small number of these cases have been reported to have responded favorably to riboflavin administration (Foy and Kondi, 1953*a*; Foy et al., 1961). Experimentally, erythroid hypoplasia with reticulocytopenia has been induced in baboons and humans by the administration of a riboflavin-deficient diet along with a riboflavin antagonist, galactoflavin (Lane et al., 1964; Lane and Alfrey, 1965; Foy et al., 1972). It should be noted, however, that the hematologic picture produced experimentally was that of erythroid hypoplasia with a shift of the erythroid cell differentiation to the left and not truly a severe erythroid aplasia. All patients on the riboflavin-deficient diet and glactoflavin after two months maintained a hemoglobin level as high as 9 g/dl and the anemia in the majority of them was not corrected to normal by the administration of riboflavin (Lane and Alfrey, 1965). Even less impressive changes in the bone marrow and blood were reported in baboons fed a riboflavin-deficient diet (Foy et al., 1972). It seems that riboflavin deficiency can cause a moderate normochromic normocytic anemia with a bone marrow morphology far different from that of the classic picture of pure red cell aplasia. Whether riboflavin deficiency in kwashiorkor contributes to the erythroid aplasia described in these cases is unknown. However, isolated riboflavin deficiency as a cause of primary pure red cell aplasia has never been documented, and in the single case of PRCA successfully treated by riboflavin it is not clear whether it was idiopathic PRCA, or mepacrine-induced PRCA, that remitted either spontaneously or following discontinuation of the drug (Foy and Kondi, 1953*a*, 1953*b*). Additional attempts to treat PRCA with riboflavin have not been successful (see chapter 4).

Megaloblastic anemia secondary to B_{12} or folic acid deficiency in very advanced cases has been recognized as a cause of severe erythroid hypoplasia (Chanarin, 1976); however, the coexistence of macro-ovalocytosis and intense myeloid hyperplasia with conspicuous megaloblastic features leads to the appropriate diagnosis. Such cases of severe erythroid hypoplasia have been mainly recognized in the course of megaloblastic anemia of pregnancy secondary to folate deficiency (Levine and Hamstra, 1969; Messerschmitt et al., 1971) as well as in selected cases of pernicious anemia and malnutrition (Rullan-Ferrer et al., 1955; Pezzimenti and Lindenbaum, 1972). Erythroid aplasia following treatment of megaloblastic anemia with B_{12} has also been described in another case and the authors attempted an association of these two conditions and raised the possibility of a causal relationship (Goldstein and Pechet, 1965). Three additional cases of pernicious anemia and PRCA occurring in the same patient have been described (Hotchkiss, 1970; Zucker et al., 1974; Robins-Browne et al., 1977). The association of PRCA and pernicious anemia can be seen only as a coexistence of two autoimmune disorders in the same patient.

Pregnancy and Erythroid Aplasia

Pure red cell aplasia has been reported in three cases in association with pregnancy (Aggio and Zunini, 1977; Miyoshi et al., 1978; Lehman and Alcoff, 1982) and in a single case during the second week postpartum during the course of a hepatitislike episode (Seidenfeld et al., 1979). Earlier reports of pure red cell aplasia in pregnancy did not provide adequate documentation of isolated bone marrow erythroid aplasia as the cause of the anemia (Skikne et al., 1976). As mentioned above, the combination of pregnancy and severe malnutrition, including severe folate deficiency, may lead to severe erythroid hypoplasia (Messerschmitt et al., 1971; Pezzimenti and Lindenbaum, 1972). In all three cases of PRCA that developed during pregnancy in the absence of nutritional deficiencies, erythropoiesis recovered in a very short time after delivery of the fetus. In one of these cases the infant was delivered by caesarean section and was found to have normal, for his gestational age, hematologic values. Interestingly enough, the mother maintained a normal hematologic picture during her subsequent second pregnancy, an observation raising questions about the existence of any association between pregnancy per se and erythroid aplasia in this case. In another case of PRCA that developed during pregnancy, erythroid aplasia persisted following delivery and remitted only after splenectomy followed by prolonged and intensive immunosuppressive treatment with azathioprine. In this probably unique case the child born from the mother with PRCA was diagnosed during his first year of life as having congenital hypoplastic anemia (Diamond-Blackfan syndrome) (Del Prete S and Cornwell GG, personal communication, 1984). This rare association of acquired PRCA in the pregnant mother with development of congenital hypoplastic anemia in the child raised the possibility that these two forms of selective erythroid aplasia, although seemingly pathogenetically different, may indeed have a common cause that is expressed in a rather different manner in the adult than in the fetus. In one of our patients who developed red cell and megakaryocytic aplasia during her first pregnancy, the erythroid and megakaryocytic aplasia did not remit within the four weeks following delivery of a normal child, but very slowly and gradually improved without any treatment over one and a half years. Although PRCA is at least temporarily related to pregnancy, its course is usually, but not always, self-limited and whether there is any pathogenetic link between pregnancy and erythroid aplasia is not known.

The effects of PRCA on the course of pregnancy and the fetus are not known. A case of a mother with acquired chronic PRCA who gave birth to infants with hydrops fetalis in three successive pregnancies has been reported. The third infant treated with intrauterine red cell transfusions survived and was found to have red cell aplasia that lasted for the first three months of life. Autopsy of the two other infants showed erythroid hypoplasia. The authors postulated that antibodies to red cell marrow progenitors were transferred transplacentally and affected the erythropoiesis of the fetuses (Oie et al., 1984).

Miscellaneous Diseases and Erythroid Aplasia

PRCA has been reported in association with various diseases characterized by an abnormal immune status such as angioimmunoblastic lymphadenopathy (Mannoji et al., 1981; al-Hilali and Joyner, 1983), acquired immunodeficiency syndrome (Berner et al., 1983; Levitt et al., 1983*a*), the syndrome of multiple endocrine gland insufficiency (Myers et al., 1980), autoimmune hypothyroidism (Browman et al., 1976; Gill et al., 1977; Francis, 1982), ulcerative colitis and chronic liver disease (Fox and Firkin, 1978), and with various paraneoplastic syndromes associated with thymomas, most commonly myasthenia gravis (Weinbaum and Thomson, 1955; Roland, 1964; Hinrichs and Stevenson, 1965; Albahary, 1972, 1977; DeSevilla et al., 1975; Houghton and Toghill, 1978; Imamura et al., 1978; Bourgeois-Droin et al., 1981; Socinski et al., 1983).

PRCA has also been described in association with renal failure (Richet et al., 1954; Gasser, 1957); however, this association seems to be rare and the reported cases do not provide adequate data to exclude other secondary forms of erythroid aplasia.

2. Clinical and Laboratory Findings

Physical Findings

The majority of adult patients with PRCA present with symptoms of anemia, which at the time of diagnosis is usually quite severe. In children with congenital hypoplastic anemia or transient erythroblastopenia of childhood, the anemia is most frequently discovered during a visit to a pediatrician either for a routine follow-up or because of symptoms suggestive of an upper respiratory viral illness. Physical examination in primary PRCA is usually negative except for pallor and signs of anemia. In secondary PRCA, physical findings are commonly limited to those caused by the underlying disease. In newly diagnosed cases of primary PRCA there is no hepatomegaly, splenomegaly, or lymphadenopathy, and enlargement of one of these organs should lead to a search for a condition other than PRCA. However, patients with PRCA unresponsive to all forms of therapy, who have been chronically under supportive care with red cell transfusions, may develop mild to moderate hepatosplenomegaly. In such patients, after years of therapy with red cell transfusions iron overload develops and signs of transfusional hemosiderosis may appear. In children with congenital hypoplastic anemia the physical examination may reveal the presence of one of various congenital abnormalities (table 7) that have been described in association with this form of erythroblastopenia.

Peripheral Blood Findings

The peripheral blood smear should be examined before any red blood cell transfusion is given. In adult PRCA and in TEC the erythrocytes are normochromic and normocytic, whereas in congenital hypoplastic anemia the red cells tend to be relatively macrocytic with an average MCV of 95 fl or greater (Diamond et al., 1976). No polychromatophilic cells are found on the smear, and the reticulocyte count is low, generally being between 0

Table 7 Congenital Abnormalities in Diamond-Blackfan Syndrome

Low birth weight	Strabismus
Short stature	Cataracts
Micro or macrocephalia	Blue sclerae
Cleft palate and/or lip	Webbed neck
Flat nasal bridge	Thumb abnormalities
Microphthalmia	Abnormal pyelocalyceal system
Hypertelorism	Congenital cardiac septal defects
Epicanthic folds	Mental retardation
Ptosis	Hypogonadism

and 1 percent. In congenital hypoplastic anemia a relatively higher reticulocyte count, as high as 5 percent, is occasionally observed. In adults the presence of polychromatophilic red cells on blood smears or of a reticulocyte count above 2 percent must raise serious doubt about the correctness of the diagnosis of PRCA. Such a reticulocyte count in the presence of severe anemia, although indicative of an anemia secondary to low red cell production, at the same time suggests that there is still a degree of persisting effective erythropoiesis as seen in cases of erythroid hypoplasia associated with myelodysplastic or myeloproliferative disorders.

The white blood cell count and the differential count are commonly normal. In an occasional patient a mild leucopenia may be noted, and lymphocytosis as well as eosinophilia have been described in a number of cases (Kark, 1937; Moeschlin and Rohr, 1943; Strauss, 1943; Foy and Kondi, 1953*a*; Tsai and Levin, 1957; Hirst and Robertson, 1967; Krantz and Zaentz, 1977). The presence of granulocytopenia, shift of the leucocyte differential count to the left, or acquired Pelger-Huët anomaly in granulocytes points against the diagnosis of PRCA.

The peripheral blood platelet count is usually normal at the time of diagnosis. Mild thrombocytopenia with platelet counts between 100,000 and 150,000/μl is occasionally seen at the time of diagnosis (Hirst and Robertson, 1967; Krantz and Zaentz, 1977), and a number of patients with refractory PRCA may later develop a more severe thrombocytopenia secondary to congestive splenomegaly. The presence, however, of severe thrombocytopenia at the time of diagnosis is not a feature of PRCA. In a number of adult patients with PRCA, and more frequently in children with TEC, thrombocytosis is found with platelet counts as high as 1,000,000/μl (Toogood et al, 1978; Corrigan, 1981; Labotka et al., 1981; Dessypris et al., 1982). This thrombocytosis seems to be reactive to anemia and the platelet count always returns to normal with restoration of erythropoiesis and a rise of hematocrit to normal levels.

Bone Marrow Morphology

The hallmark of PRCA is the absence of erythroblasts from an otherwise normal marrow (figure 1). The marrow cellularity is normal or mildly increased. Marrow cellularity can be more accurately assessed on bone marrow biopsy or marrow particle sections rather than on bone marrow smears. In an occasional young patient the marrow cellularity may be increased, but in adults the marrow very seldom is 90–100 percent cellular. Such an abnormally high marrow cellularity with almost complete obliteration of fat spaces on biopsy or particle sections must raise suspicion about the presence of a myelodysplastic or myeloproliferative disorder. In a classic case of PRCA in bone marrow cell smears no erythroblasts, or only very few orthochromatic and/or polychromatophilic erythroblasts, are seen, constituting less than 0.5–1 percent on the differential bone marrow cell count. In bone marrow biopsy, erythroblasts are rare and only a single erythroblastic island consisting of a few late erythroblasts may be found. In the majority of cases the rare erythroblasts seen are morphologically normal but occasionally they might show nuclear dysplasia or intracytoplasmic vacuoles. The presence of small numbers of vacuolated giant proerythroblasts has been associated with erythroid aplasia seen in severe nutritional deficiency (Kho, 1957; Kondi and Foy, 1964; Zucker et al., 1971). In a small number of cases a few early erythroblasts have been described (Eisenmann and Dameshek, 1954; Gasser, 1957; Krantz and Kao, 1969*b*; Wranne, 1970; Bottiger and Rausing, 1972; Zaentz et al., 1975). In these cases of PRCA proerythroblasts and early basophilic erythroblasts may constitute up to 5 percent of the nucleated cells in the bone marrow differential cell count. In adult cases of chronic PRCA, the presence of these early erythroid cells in the marrow should not be interpreted as evidence of recovering erythropoiesis, since in these patients the disease has the same clinical course as cases characterized by complete erythroid aplasia. In contrast, in TEC such a finding represents an early recovery phase of erythropoiesis which becomes apparent within a few days by a rise of reticulocyte count in the peripheral blood. In congenital hypoplastic anemia, particularly during the first few months of life, the degree of erythroid hypoplasia is not as severe as in later life, and erythroid cells in the marrow may constitute as much as 10 percent of marrow cells (Diamond et al., 1976). In addition, in 5 to 10 percent of cases erythroid hyperplasia or normal numbers of erythroblasts with maturation arrest at the stage of proerythroblast and/or megaloblastic features can be seen (Arrowsmith et al., 1953; Bernard et al., 1962).

In a limited number of cases a period of ineffective erythropoiesis characterized by erythroid hyperplasia with maturation arrest at the stage of proerythroblasts or basophilic erythroblasts in the marrow and reticulo-

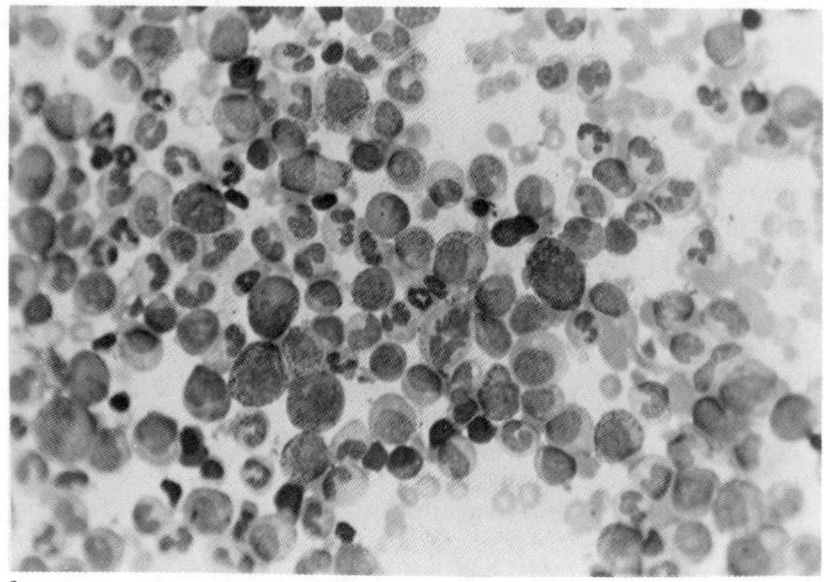

a

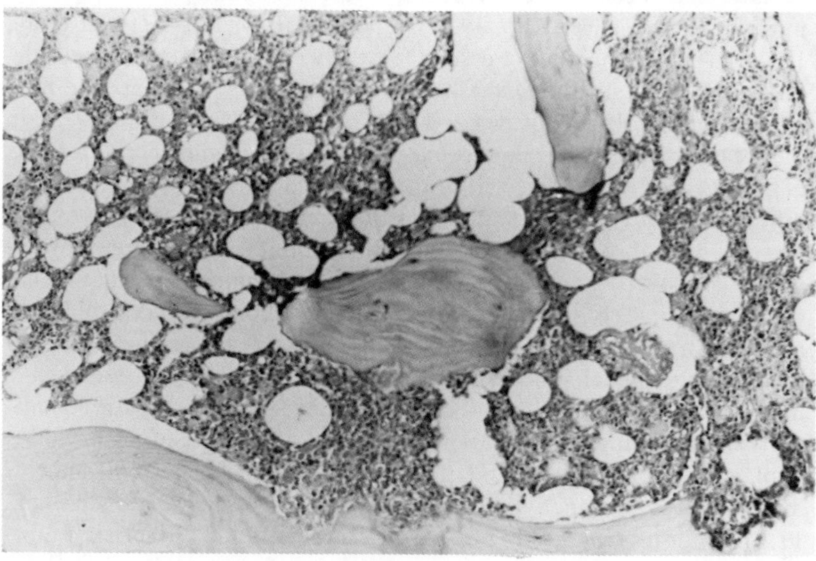

b

Figure 1 (*A*) Bone marrow aspirate from a patient with pure red cell aplasia showing a predominance of myeloid cells, a mild increase in the number of small lymphocytes, and an almost complete absence of erythroid precursors. (*B*) Bone marrow biopsy from the same patient showing normal cellularity with normal myeloid cells and megakaryocytes and an absence of erythroblasts.

cytopenia with anemia in the peripheral blood preceded the development of PRCA, developed during its course, or developed after partial response to treatment and before erythropoiesis returned to normal (Vilter et al., 1960; Beard et al., 1978; Wibulyachainunt et al., 1978; Dessypris et al., 1984; Jacobs et al., 1985). Although such a bone marrow finding is not diagnostic of PRCA, in the absence of any other abnormality it should be considered as a phase in the natural course of this syndrome and cases with such an isolated finding in the erythroid line should be managed as cases of PRCA. In two cases a period of ineffective erythropoiesis without maturation arrest (moderate anemia, reticulocytopenia, and unremarkable bone marrow morphology) has been noted in patients previously treated successfully for their PRCA during a 3 to 4 month period preceding a relapse (Krantz and Dessypris, unpublished observations). In previously diagnosed and successfully treated cases of PRCA the development of unexplained, slowly progressing, moderate, normochromic normocytic anemia with reticulocytopenia and a normal bone marrow morphology should alert the physician about a possible relapse.

The myeloid cells and the megakaryocytes in the marrow are normal and exhibit full maturation. Shift of the granulocytic cells in the marrow to the left, presence of hypogranular myelocytes, features of megaloblastic maturation such as giant metamyelocytes, and the presence of mononuclear or dysplastic megakaryocytes are not findings consistent with the diagnosis of pure red cell aplasia.

An increased number of lymphocytes on smear or lymphoid aggregates on marrow biopsy, an increased number of plasma cells, mast cells, or eosinophils may be seen (Hirst and Robertson, 1967; Marmont et al., 1975; Zaentz et al., 1975; Dessypris and Krantz, unpublished observations). Iron stores are increased by iron stain and occasionally iron granules can be found in plasma cells. During the period of recovery and in the presence of erythroid hyperplasia with ineffective erythropoiesis an occasional ring sideroblast can be seen.

Diagnosis and Differential Diagnosis

The diagnosis of PRCA is made on the basis of the following blood and bone marrow findings: reticulocytopenia with normochromic normocytic anemia with normal leucocyte and platelet counts with a bone marrow showing isolated erythroid aplasia with normal cellularity and normal myeloid and megakaryocytic maturation and morphology. It should be emphasized that a bone marrow biopsy, preferably stained by periodic acid-Schiff (PAS), is of great help in assessing bone marrow cellularity and in confirming the severity of erythroid aplasia. Staining of the marrow biopsy by PAS allows better demonstration of nuclear morphology and thus makes the distinction between late erythroblasts and lymphocytes easier. In a small number of cases the degree of erythroid hypoplasia might be overestimated on the bone marrow smear since normally the erythroid cells are not spread as easily as the myeloid cells and provide a normal M:E ratio of 1.5:1 to 3.5:1, whereas in the biopsy the M:E ratio is normally very close to 1:1.

In adults the other hematologic syndrome from which PRCA must be differentiated is the myelodysplastic or preleukemic syndrome and unclassified, atypical myeloproliferative disorder. In the majority of cases of myelodysplastic syndrome erythroid hyperplasia rather than severe hypoplasia is observed and the marrow cellularity may be normal or increased (Pierre, 1974; Linman and Bagby, 1976; Bennett et al., 1982). In addition morphologic signs of abnormal maturation or dysplasia are frequently observed either in the erythroid or in the other cell lines. The presence in the peripheral blood smear of monocytosis, acquired Pelger-Huët anomaly, or shift of the white cell differential count to the left and the detection in the bone marrow of hypogranular forms of myelocytes, shift of the myeloid maturation to the left, collections of monocytoid blasts, or clusters of dysplastic mononuclear megakaryocytes should lead away from the diagnosis of PRCA. Even in cases of erythroid hypoplasia many foci of erythroid cells can be easily seen in the bone marrow biopsy thus excluding the diagnosis of PRCA (Dessypris and Krantz, 1985*b*). In cases where the morphologic findings in the bone marrow are not adequate to differentiate PRCA from myelodysplastic syndrome, determination of leukocyte alkaline phosphatase, cytogenetic studies on marrow cells, and assay of marrow cells for granulocytic-monocytic colony-forming units (CFU-GM) in vitro may be helpful. It should be noted that in the differential diagnosis of PRCA from the myelodysplastic syndromes the assay of marrow cells for CFU-GM is much more helpful than the assay for erythroid progenitors (CFU-E, BFU-E, see chapter 3). A low leukocyte alkaline phosphatase, the presence of an abnormal bone marrow karyotype, and subnormal or absent growth of granulocytic-monocytic progenitors (CFU-GM) in vitro are findings favoring the diagnosis of a myelodysplastic syndrome rather than idiopathic PRCA (Greenberg et al., 1971; Pierre, 1975; Linman and Bagby, 1976; Koefflffler and Golde, 1980).

In certain cases the presence of multiple lymphoid aggregates on bone marrow biopsy, or occasionally of a moderately severe diffuse lymphocytic infiltration, may raise the question of a coexisting lymphoproliferative disorder. In such cases immunologic typing of marrow lymphocytes may be helpful in solving this problem, since monoclonality of the lymphoid cell population favors the presence of a coexisting lymphoid malignancy, whereas demonstration of a polyclonal lymphoid cell population favors the presence of a reactive benign lymphoproliferation.

In a number of cases PRCA should also be differentiated from chronic forms of mild to moderately severe aplastic anemia in which anemia is the predominant hematologic finding. In these cases there is always a coexisting neutropenia or thrombocytopenia of varying severity. The bone marrow is hypocellular (usually 30 percent or less) and although erythroid hypoplasia may be the most prominent feature, there are more than 2–5 percent erythroblasts on marrow cell differential, representing all stages of erythroid differentiation. On bone marrow biopsy the presence of hypocellularity and easily identifiable small collections of erythroblasts at various stages of maturation are the two most helpful findings differentiating the PRCA from aplastic anemia.

In children the differential diagnosis of anemia, reticulocytopenia, and isolated erythroid aplasia centers between congenital hypoplastic anemia and transient erythroblastopenia of childhood. The appearance of anemia before the first year of life (figures 2 and 3), the presence of congenital anomalies on physical or radiographic examination (see table 7), the absence of recent history of respiratory or gastrointestinal "viral" illness, an erythrocyte MCV at diagnosis higher than 95 or 100 fl (figures 4 and 5), an increased percentage of hemoglobin-F for the child's age, and the presence of i antigen on red cells at the time of diagnosis strongly favor the diagnosis of congenital hypoplastic anemia. In this condition it seems that erythropoiesis maintains its fetal characteristics, as expressed by an increased amount of i antigen commonly present on cord blood erythrocytes, increased fetal hemoglobin, and increased levels of various glycolytic enzymes (Wang and Mentzer, 1976; Alter, 1980). The distribution of fetal hemoglobin among red cells is uneven, and the ratio of glycine to alanine in the 136 position of the beta chain favors glycine, indicating that the type of Hb-F present in erythrocytes is of the fetal and not the adult type (Schroeder et al., 1971; Alter, 1980). This fetallike erythropoiesis persists during recovery and during phases of remission. These laboratory findings are quite helpful in distinguishing the congenital from the acquired form of erythroid aplasia of childhood. However, it should be noted that these tests must be performed at the time of diagnosis and are valid only when red cells are examined at a time distant from the recovery phase of TEC, during which a phase of fetallike erythropoiesis is observed, which in contrast to CHA disappears when the hematologic values return to normal at full recovery (Papayannopoulou et al., 1980; Link and Alter, 1981).

Cytogenetic Studies

Cytogenetic studies on bone marrow cells in chronic primary PRCA reveal in the great majority of cases a normal karyotype. In a group of 31 patients studied at Vanderbilt University between 1970 and 1980 and in an additional group of 8 patients seen between 1980 and 1985 only one patient revealed an abnormal marrow cell karyotype (47, XY, +G) (Dessypris et

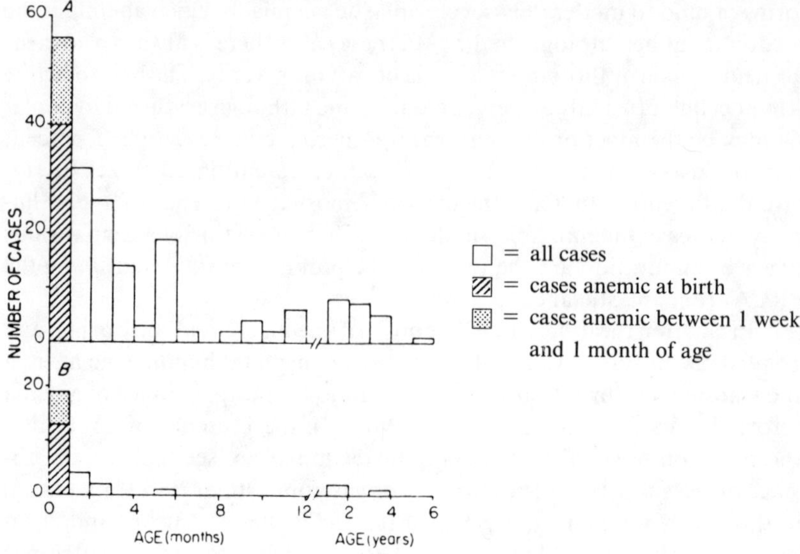

Figure 2 Age at diagnosis in patients with congenital hypoplastic anemia. (*A*) 186 cases reported in the literature; (*B*) 29 cases seen at Boston's Children's Hospital Medical Center. (From Alter, 1980.)

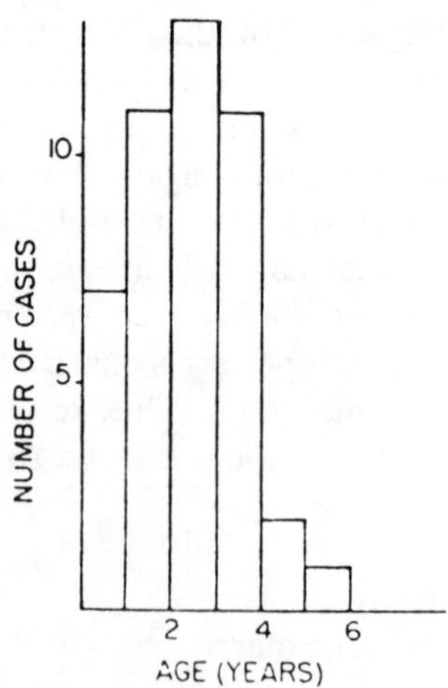

Figure 3 Age at diagnosis in 45 patients with transient erythroblastopenia of childhood. (From Alter, 1980.)

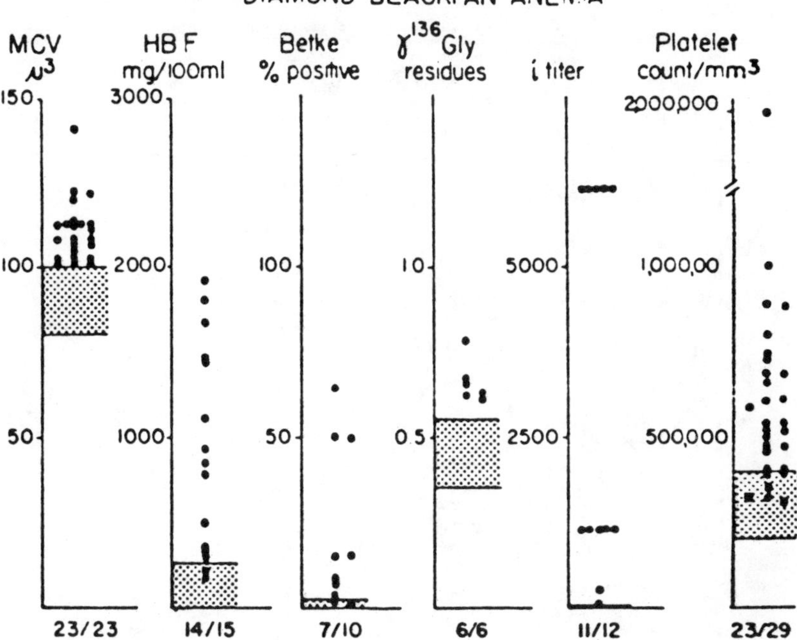

Figure 4 Laboratory findings in treated patients with congenital hypoplastic anemia. The vertical axis shows units for each parameter. The numbers at the bottom represent the number of cases with abnormal values as related to the total number of tested cases. The shaded area corresponds to the normal range (± 2 SD). (From Alter, 1980.)

al., 1980). Two more cases of erythroid hypoplasia have been described with deletion of the long arm of a B-group chromosome in one (Fitzgerald and Hamer, 1971), and with the 5 q(-) abnormality in the other (DiBenedetto et al., 1979). In another case of PRCA the Philadelphia chromosome (Ph^1) was detected in metaphases of the patient's bone marrow cells at a time when there were no hematologic findings of chronic granulocytic leukemia. This patient responded to treatment with prednisone and, despite an increase in the percentage of the Ph^1 positive cells in the bone marrow during remission, maintained normal blood and bone marrow morphology for at least a year following the diagnosis of PRCA (Shiraishi et al., 1980). The detection of an abnormal clone in the patient's marrow cells by cytogenetic analysis has been suggested to have prognostic significance and it has been considered as an early sign of leukemic transformation (preleukemic PRCA) (Dessypris et al., 1980).

In congenital hypoplastic anemia, peripheral blood chromosomes are essentially normal (Alter, 1980), a finding differentiating this condition

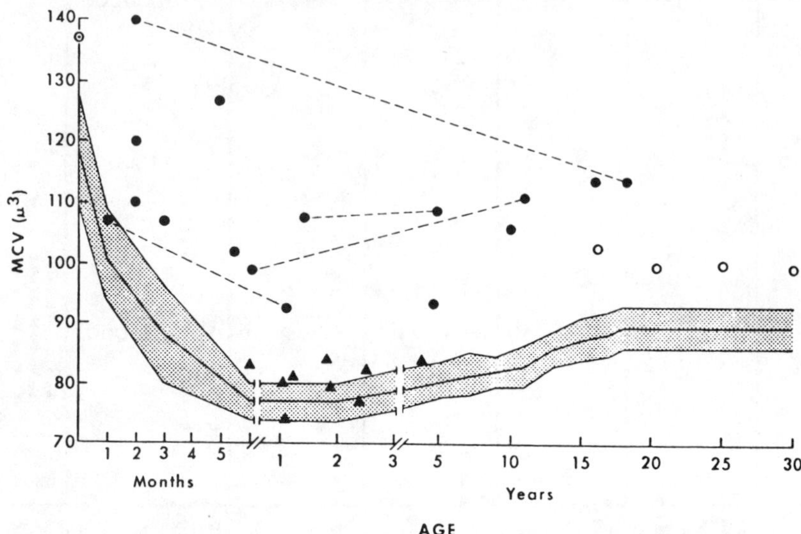

Figure 5 The mean corpuscular volume (MCV) of erythrocytes in hypoplastic anemia of childhood. The shaded area represents the normal average MCV ± 1 SD. Intermittent lines represent the MCV of the erythrocytes of the same patient at different ages after treatment with corticosteroids. (Modified from Wang and Mentzer, 1976.)

from Fanconi's anemia, where an increased incidence of chromosomal breaks has been noted (Bloom et al., 1966). In a few cases various abnormalities have been described such as an achromatic gap, or a pericentric inversion of chromosome 1, and an enlarged chromosome 16 (Khatua, 1964; Tartaglia et al., 1966; Heyn et al, 1974). The sister chromatid exchange rate has been reported to be normal in six studied cases (Alter, 1980). A patient with an increased rate of spontaneous and x-ray-induced dicentric chromosomes and micronuclei, but with a normal frequency of sister chromatid exchange, has also been reported (Iskandar et al., 1980). Bone marrow cytogenetic studies have not been reported in congenital hypoplastic anemia or transient erythroblastopenia of childhood.

Ferrokinetic Studies

During the initial phase of PRCA, ferrokinetic studies show a markedly prolonged clearance of ^{59}Fe, no significant accumulation of ^{59}Fe

in the sacral bone, and undetectable or extremely low red cell ^{59}Fe incorporation within the two-week period following the injection of radioisotope (Pollycove and Mortimer, 1961; DiGulielmo et al., 1963; Schmid et al., 1963; Finch et al., 1970; Marmount et al., 1975; Pollycove and Tono, 1975; Prasad et al., 1976; Viala et al., 1981). Similar findings have also been reported in children with congenital hypoplastic anemia (Bernard et al., 1962*b*; Tartaglia et al., 1966; Sjölin and Wranne, 1970; Falter and Robinson, 1972; Lawton et al., 1974). These findings are not specific for isolated erythroid marrow aplasia but are common in all conditions in which there is little or no residual erythropoiesis in the marrow regardless of its histology (Finch et al., 1970). After remission of the disease all the abnormal ferrokinetic parameters return to normal (Böttiger and Rausing, 1972; Marmont et al., 1975; Wibulyachainunt et al., 1978; Viala et al., 1981). In an occasional patient, however, findings consistent with ineffective erythropoiesis may be present, such as a very short ^{59}Fe clearance rate and reduced red cell ^{59}Fe incorporation associated with a marked increase of the erythron iron turnover over the fixed red cell iron turnover and a prolonged marrow transit time (Wibulyachainunt et al., 1978). This condition is usually associated with moderate anemia with a low reticulocyte count and erythroid hyperplasia in the bone marrow with maturation arrest at the level of early basophilic erythroblast. Such a state of ineffective erythropoiesis may precede the development or follow treatment of PRCA with immunosuppressive agents (Beard et al., 1978; Wilbulyachainunt et al., 1978).

Bone Marrow Imaging

Bone marrow scanning by the use of ^{59}Fe or ^{111}In shows little or no uptake of the radioisotope by the bone marrow (Bunn et al., 1976). In a report by Merrick et al. (1975), however, extensive bone marrow uptake of ^{111}In was noted in six patients with red cell aplasia. Five of these patients had coexisting neoplastic disorders and no hematologic data were reported to confirm the diagnosis of PRCA. The uptake of ^{59}Fe by the bone marrow returns to normal after remission of PRCA, but in a number of cases a significant expansion of erythropoietic activity toward the proximal portions of the lower extremities has been noted (Wibulyachainunt et al., 1978).

Red Cell Survival

In both children and adults with isolated erythroid aplasia, the survival of red cells tagged with ^{51}Cr has been found to be mildly shorter than normal (Schmid et al., 1963; Wibulyachainunt et al., 1978; Feldges and

Schmidt, 1971; Alter, 1980). Previous studies of red cell survival by the Ashby technique have demonstrated a normal erythrocyte life span (Tsai and Levin, 1957). The decreased survival of autologous or compatible transfused red cells is always inadequate to explain the severity of the anemia except in those cases in which PRCA develops during the course of congenital or acquired hemolytic anemias. In chronic cases of PRCA refractory to treatment after many years of red cell transfusions, red cell survival may decrease significantly either because of development of red cell alloantibodies, or because of the development of splenomegaly with work hypertrophy. This is usually manifested clinically by an increase in the red cell transfusion requirements.

Erythrocytic Enzymes

The activity of various red cell enzymes has been studied by Wang and Mentzer (1976) in 11 children with congenital hypoplastic anemia and in 9 children with transient erythroblastopenia of childhood (figure 6). Significant differences were noted in the activities of various red cell enzymes between the congenital and the acquired forms of red cell aplasia in children. Among the various red cell enzymes glyceraldehyde-3-phosphate dehydrogenase, lactate dehydrogenase, transaminases, aldolase, phosphofructokinase, and glutathione peroxidase were found to be particularly useful in differentiating the congenital from the acquired red cell aplasia in children. In congenital hypoplastic anemia, the activity of these red cell enzymes was found to be above the normal mean and even greater than the mean activity in reticulocyte-rich blood. In contrast, the activity of these enzymes in red cells from children with TEC was below the normal mean, indicating the presence of a population of old erythrocytes in the blood of these children (Sass et al., 1964). Similar studies performed in another group of five patients with TEC by Tillmann et al. (1976) confirmed these findings. The detection of high activity of certain red cell enzymes in the blood of patients with congenital hypoplastic anemia represents another manifestation of the fetallike erythropoiesis present in this disorder. In two patients with congenital hypoplastic anemia in continuous remission for 3 and 15 years the level of activity of red cell enzymes was found to be very close to normal (Wang and Mentzer, 1976).

Immunologic Abnormalities

A variety of abnormalities of the immune system have been reported in patients with chronic acquired PRCA including hypogammaglobulinemia (Ramos and Loeb, 1956; Linsk and Murray, 1961; DiGiacomo et al., 1966; Murray and Webb, 1966; Dameshek et al., 1967; Hanzawa et al., 1969; Souquet et al., 1970; Vasavada et al., 1973; Geary et al., 1975; Kozuru et al., 1977), paraproteinemia or monoclonal gammopathy (Voyce, 1963; Krantz and Kao, 1969; Lemenager et al., 1973; Resegotti and Ricci, 1976; Resegotti et al., 1978; Varet et al., 1978), pyroproteins (Krantz and Kao, 1969), decreased complement levels (Krantz, 1974), anergy (Krantz and Zaentz, 1977), antinuclear or antismooth muscle cell antibodies, or a positive lupus erythematosus cell preparation (Holborow et al., 1963; Barnes, 1965; River, 1966; Krantz and Kao, 1967, 1969; Vilan et al., 1973; Sao et al., 1982), and autoimmune hemolytic anemia (Eisemann and Dameshek, 1954; Gasser, 1957; DiGiacomo et al., 1966; Albahary, 1977; Meyer et al., 1978; Clauvel et al., 1983; Mangan et al., 1984; Dessypris and Krantz, unpublished observations). It should be emphasized, however, that a number, but not all, of reported cases with the above immunologic abnormalities were associated with thymoma, and whether these abnormalities of the immune system are related to the coexisting thymoma or the PRCA per se is not clear. As has been pointed out in a single patient with PRCA, only one or two of the above abnormalities may be present, but in the whole group of patients with chronic erythroid aplasia there have been various clinical or laboratory abnormalities associated with autoimmune disease (Krantz and Zaentz, 1977) (table 8).

Abnormalities in the peripheral blood lymphocytes have been reported in patients with PRCA. Decreased number of B-lymphocytes have been reported in four patients. One of them had an abnormally high percentage of lymphocytes with Fc receptors, which became normal 3 years after the initial study. Normal lymphocyte transformation by various mitogens was reported in three out of four studied patients. In two cases a decrease in the phytohemagglutinin-induced lymphocyte-mediated cytotoxicity was noted (Froland et al., 1976). In a small number of cases of what was thought to be acquired idiopathic PRCA an increased number of large granular lymphocytes bearing receptors for the Fc portion of IgG(T_γ) have been reported (Linch et al., 1981; Abkowitz et al., 1986). In one of these cases, T cells were also abnormally increased in the marrow of the patient and whether this case was idiopathic or secondary to T cell chronic lymphocytic leukemia (or lymphoproliferative disorder) is not clear. In a case of PRCA associated with thymoma and hypogammaglobulinemia the patient's peripheral blood T cells and thymus cells did not express the epitope recognized by OKT4A and Leu3A monoclonal antibodies. This patient had intact delayed hypersensitivity and his peripheral lymphocytes proliferated normally on exposure to various lectins and antigens. The increased suppressor activity that was associated with a relative increase in OKT8 lymphocytes was abrogated by irradiation, which restored T-helper activ-

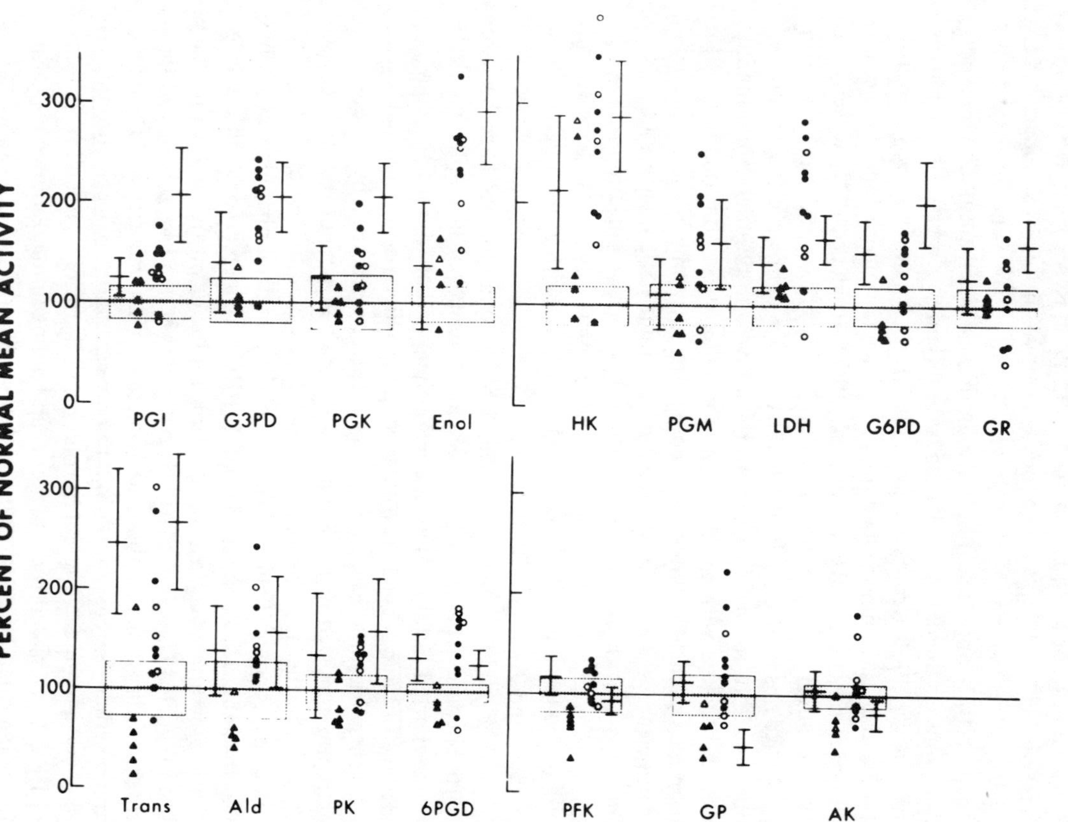
PERCENT OF NORMAL MEAN ACTIVITY
300
200
100
0
PGI
G3PD
PGK
Enol
HK
PGM
LDH
G6PD
GR
Trans
Ald
PK
6PGD
PFK
GP
AK

Figure 6 The activity of various enzymes in the erythrocytes of children with hypoplastic anemia. TEC erythrocytes had normal or low activity, except during recovery, when the reticulocyte count was high. CHA erythrocytes had above-normal activity. The shaded areas represent the mean ± 1 SD for normal children and adults. (From Wang and Mentzer, 1976.)

+ = the mean ± 1 SD for reticulocyte-rich blood (*left*) and for cord blood (*right*)

▲ = transient erythroblastopenia of childhood

△ = transient erythroblastopenia of childhood during recovery

● = congenital hypoplastic anemia on therapy

o = congenital hypoplastic anemia off therapy

AK = adenylate kinase

Ald = aldolase

Enol = enolase

G3PD = glyceraldehyde-3-phosphate dehydrogenase

G6PD = glucose-6-phosphate dehydrogenase

GP = glutathione peroxidase

GR = glutathione reductase

HK = hexokinase

LDH = lactate dehydrogenase

PGI = phosphoglucoseisomerase

PGK = phosphoglycerokinase

PFK = phosphofructokinase

PGM = phosphoglyceromutase

PK = pyruvate kinase

Trans = transaminase

6PGD = 6-phosphogluconate dehydrogenase

Table 8 Abnormalities and Diseases of the Immune System Associated with Pure Red Cell Aplasia

Diseases
Thymoma
Myasthenia gravis
Lymphoma
Chronic lymphocytic leukemia
Angioimmunoblastic lymphadenopathy
Acquired immunodeficiency syndrome
Autoimmune hemolytic anemia
Systemic lupus erythematosus
Adult and juvenile rheumatoid arthritis
Autoimmune hypothyroidism
Syndrome of multiple endocrine gland insufficiency
Pernicious anemia
Laboratory abnormalities
Hypogammaglobulinemia
Anergy
Positive antiglobulin test
Monoclonal gammopathy
Pyroproteins
LE cell phenomenon
Antinuclear antibodies
Anti-DNA antibodies
Antismooth muscle antibodies
Antithyroid antibodies
Antiintrinsic factor antibodies
Decreased complement

ity for B cell polyclonal proliferation to normal (Levinson et al., 1985). The significance of these findings in relation to the pathogenesis of erythroid aplasia remains unclear.

In a number of primary or secondary cases of PRCA, a decreased ratio of OKT4:OKT8 (inducer-helper to suppressor-cytotoxic) lymphocytes, due to the presence of an increased number of OKT8 lymphocytes, has been reported. This abnormality may disappear after the disease goes into remission following treatment (Kuriyama et al., 1984; Milnes et al., 1984; Berlin and Lieden, 1986). The significance of these findings and their association with the pathogenesis of PRCA are not clear since changes in the percentage of either helper or suppressor lymphocytes in the peripheral blood resulting in an abnormal helper to suppressor ratio (OKT4:OKT8) have been reported in a variety of conditions including aplastic anemia (Kuriyama et al., 1984; Zoumbos et al., 1984), myelodysplastic syndromes (Bynoe et al., 1983), and in a number of patients who have received multiple red cell transfusions (Gascon et al., 1984). Multiply transfused patients with idiopathic PRCA are expected to have an increased number of T-lymphocytes expressing the HLA-DR antigen (activated lymphocytes) and a decreased natural killer cell function as it has been described in a number of multiply transfused patients with a variety of hematologic disorders (Gascon et al., 1984; Kaplan et al., 1984).

In a small number of children (10) with congenital hypoplastic anemia, hypogammaglobulinemia has been reported (Bernard et al., 1962*b*; Brookfield and Singh, 1974; Alter 1980). An increase in the T_{γ} -lymphocytes in the blood and bone marrow has been described in a single child with congenital hypoplastic anemia (Cornaglia-Ferraris et al., 1981). In four of five children with CHA, a depressed OKT4:OKT8 ratio has been reported. The same group of patients has been shown to demonstrate an inability to generate suppressor cells following stimulation by concanavalin-A and impaired prostaglandin-mediated suppression of lymphocyte proliferation (Finlay et al., 1982). These investigators concluded that in CHA the defect is not restricted to the erythroid progenitor cells but extends also to the lymphocytes. Further studies are necessary to confirm these findings and the extent to which this conclusion is correct.

Radiologic Findings

In adults with primary PRCA, abnormal radiologic findings may be confined to the chest. Since there is a significant association of thymomas with PRCA a search for a thymoma is indicated in any newly diagnosed case of primary PRCA. Today it seems that the most sensitive method in detecting a thymoma in a patient with a normal chest radiograph is the computer-assisted tomography of the anterior mediastinum. With this technique a thymoma can be easily diagnosed in patients over 40 years of age. In younger patients, however, this technique cannot differentiate a thymoma from normal thymus or thymic hyperplasia with a high level of confidence, but it remains the most sensitive method in excluding the presence of a thymoma (McLoud et al., 1979; Baron et al., 1982; Fon et al., 1982). The use of computerized tomography of the mediastinum may lead to detection of a mediastinal mass that cannot be suspected on the plain chest radiograph or on lateral tomograms of the mediastinum; therefore, a workup for thymomas cannot be considered complete without a computerized tomogram of the mediastinum. It should be noted, however, that small thymomas, less than 0.5 cm in diameter, may be missed even with this technique. In a limited number of cases, preoperative visualization of thymomas has been achieved with ^{75}Se-selenomethionine or by ^{67}Ga-gallium citrate scanning. The sensitivity and specificity of these radionuclear scanning techniques for the diagnosis of small thymomas is not known (Toole and Witcofski, 1966; Hare and Andrews, 1970; Cowan et al., 1971; Higasi et al., 1972; Min et al., 1978).

In children, thymomas as well as chronic acquired PRCA are extremely rare, and detailed searches for thymomas are not thought to be necessary. In children with congenital PRCA, radiologic findings may be limited to those associated with coexisting congenital abnormalities (see table 7).

Miscellaneous Laboratory Abnormalities

Serum iron, ferritin, and percent of saturated transferrin are increased due to underproduction of red cells and the transfusions of erythrocytes. Serum folate and B_{12} levels are within the normal limits. Serum haptoglobin is generally normal, unless this test is performed less than a week to ten days following a red cell transfusion or the erythroid aplasia is associated with a congenital or acquired hemolytic anemia. Many patients, particularly those with a history of multiple transfusions, have mild elevation of transaminases, alkaline phosphatase, and lactic dehydrogenase in the serum, probably reflecting a form of an indolent transfusion-associated hepatitis. In cases refractory to all forms of treatment and totally dependent on red cell transfusions, iron overload eventually develops with all the associated laboratory findings of liver and endocrine gland dysfunction.

3. Pathogenesis

During the last 35 years, the pathogenesis of erythroid marrow aplasia has been the subject of intensive investigation. In 1949, Smith described a case of chronic congenital red cell aplasia in a newborn infant isoimmunized to blood group antigen A by his mother and hypothesized that this might have been the cause of the erythroid aplasia (Smith, 1949). The association of pure red cell aplasia with thymomas and with autoimmune hemolytic anemia has been subsequently recognized and the hypothesis was raised that antibodies against the red cells may also attack the erythroid precursors in the marrow and cause erythroid aplasia (Gasser, 1949; Rohr, 1949; Eisenmann and Dameshek, 1954; Seip, 1955). Based on this hypothesis, Loeb attempted to induce erythroid aplasia in a normal human volunteer by injecting him intravenously with 250 ml of fresh plasma collected from a unique patient with thymoma associated with hypogammaglobulinemia, pure red cell aplasia, and extramedullary splenic hematopoiesis (Loeb, 1956). This human experiment was not successful. In the following years, various in vivo and in vitro methods for the study of erythropoiesis were developed. The application of these techniques for the investigation of the pathogenesis of pure red cell aplasia has provided an increasing amount of evidence in favor of its immune pathogenesis.

The Effect of PRCA Serum on the Erythropoiesis of Animals in Vivo

The first studies on the pathogenesis of red cell aplasia were performed on rodents or rabbits by injecting them with serum or plasma from PRCA patients and assaying its effect on the rate of red cell production as measured by the amount of ^{59}Fe incorporated into red cells in the peripheral blood of the animals. In 1964, Entwistle and associates studied the effect of serum from a patient with PRCA and carcinoma of the bronchus on the rate of red cell production by normal rabbits (Entwistle et al., 1964). In an attempt to demonstrate a humoral factor capable of depressing red cell production in experimental animals, they injected 1 ml of serum into rabbits intraperitoneally daily for five days starting one day before the injection of ^{59}Fe and ten days later they measured the amount of ^{59}Fe radioactivity in the red cell mass of the animals. In these experiments they demonstrated that the mean iron utilization in animals injected with normal serum was 50.4 percent, whereas in animals injected with PRCA serum mean iron utilization dropped to 1.2 percent. Serum collected and tested after irradiation of the bronchial tumor had no effect on iron incorporation into the red cells of rabbits. They concluded that the patient's serum contained a humoral factor inhibiting iron utilization by the rabbit and they suggested that an antibody to erythropoietin may be able to produce red cell aplasia in humans in a fashion similar to the reduction of erythropoiesis in mice injected with antierythropoietin antibodies (Schooley and Garcia, 1962, 1965).

Using the exhypoxic-polycythemic mouse assay Jepson and Lowenstein (1966) studied the effect of plasma from two patients with PRCA on murine erythropoiesis in vivo. When mice are exposed to a 10 percent oxygen atmosphere they increase red cell production, which leads to an increase in their red cell mass and development of polycythemia. This polycythemia is induced by increased levels of erythropoietin circulating in the blood of these animals as a normal response to hypoxia. Upon return of these animals to normal atmosphere—containing 21 percent oxygen—the stimulus of hypoxia is removed and endogenous erythropoietin release is almost totally inhibited, resulting in cessation of red cell production which lasts until the hematocrit returns to normal. During this period of experimentally suppressed erythropoiesis, injection of the animals with erythropoietin preparations leads to stimulation of red cell production which can be quantified by measuring the amount of ^{59}Fe incorporated by the newly formed red blood cells released in the blood during the 48-hour period following injection of erythropoietin. This system is called the ex-hypoxic-polycythemic mouse assay and has been used as a method for assaying the erythropoietic activity of various substances, including human plasma and urine, from patients with disorders of erythropoiesis (Boivin and Eoche-Duval, 1965; Krantz and Jacobson, 1970). When 0.5 ml of plasma from a patient with PRCA associated with thymoma and an equal amount of plasma from a patient with acute erythroblastopenia were injected into polycythemic mice along with a urinary extract containing erythropoietin, a 33 to 85 percent inhibition of ^{59}Fe incorporation into red cells was noted. Normal plasma, however, as well as plasma from patients with erythroleukemia, chloramphenicol-induced aplastic anemia, or idiopathic aplastic anemia, had no significant effect on the rate of erythropoietin-induced red cell production by the polycythemic mice. The investigators concluded that the plasmas of the two patients contained an erythropoietic inhibitor with biologic activity similar to that in antisera raised against erythropoietin. They postulated that the inhibitory effect of these plasmas may be due to the presence of an antibody directed against erythropoietin (Jepson and Lowenstein, 1966).

Another patient with thymoma and PRCA was studied by similar methods by al-Mondhiry and colleagues (1971). The investigators showed that the patient's plasma contained increased levels of erythropoietin that could be completely neutralized by an antierythropoietin antiserum. The serum and urine of this patient collected before thymectomy inhibited ^{59}Fe incorporation into red cells in the polycythemic mouse assay and this inhibitory activity was not detectable in serum collected after thymectomy and recovery of erythropoiesis. The inhibitor was present in the patient's serum IgG fraction which failed to inactivate exogenous erythropoietin after prior neutralization of the endogenous erythropoietin with antierythropoietin antiserum. The fact that the urine of this patient also contained an inhibitory activity that could not be an IgG has not been addressed. The investigators concluded that the inhibitor was not directed against erythropoietin but most likely against the erythroid marrow cells.

In a series of ten patients with erthroblastopenia studied by the polycythemic mouse assay, Jepson and Vas (1974) were able to demonstrate the presence of such an inhibitor in the IgG fraction of the serum of eight of them. The plasma of these patients contained high levels of erythropoietin. The inhibition of ^{59}Fe incorporation into the polycythemic mouse red cells was found to increase with injections of increasing amounts of PRCA serum IgG whereas normal IgG had no effect (figure 7). These investigators suggested that the IgG in the serum of these patients damaged erythroblasts much as serum antibodies injure circulating erythrocytes in autoimmune hemolytic anemia, leading to the clinical picture of erythroblastopenia. However, direct experimental evidence for this conclusion was not provided. A patient with erythroblastopenia that developed two years after thymectomy was also studied by Zalusky et al. (1973). The serum and its IgG fraction suppressed erythropoiesis in normal mice. Addition of excess of erythropoietin did not overcome this inhibition. Chronic injections of this patient's serum into mice resulted in reticulocytopenia, anemia, and suppression of the 48-hour ^{59}Fe incorporation into erythrocytes, despite a concurrent rise in the plasma erythropoietin levels in these animals. Normal rat bone marrow absorbed the inhibitor, but hypertransfused rat bone marrow, which is depleted of erythroid precursors, did not. These investigators concluded that the inhibitor interferes with erythropoiesis by acting directly on the early erythroid cells.

Peschle and co-workers (1978) also studied the effect of PRCA serum IgG on the ^{59}Fe incorporation by red cells of normal mice and on the number of erythroid progenitors, colony-forming units (CFU-E) per tibia

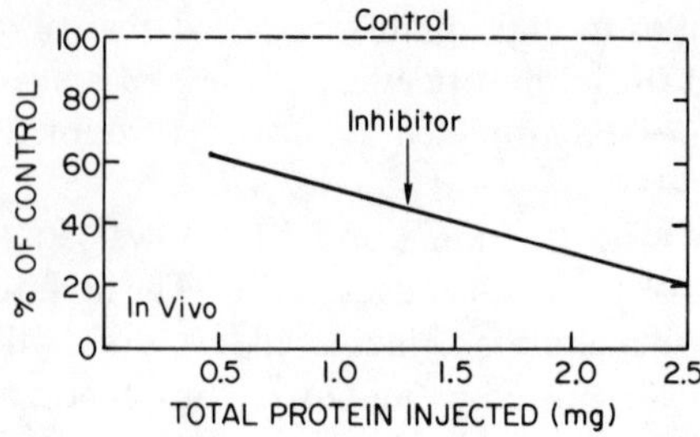

Figure 7 The effect of plasma IgG from a patient with pure red cell aplasia on the incorporation of ^{59}Fe into red cells in exhypoxic polycythemic mice. IgG was injected intraperitoneally simultaneously with erythropoietin, following it, and 24 hours later. (From Jepson and Vas, 1974.)

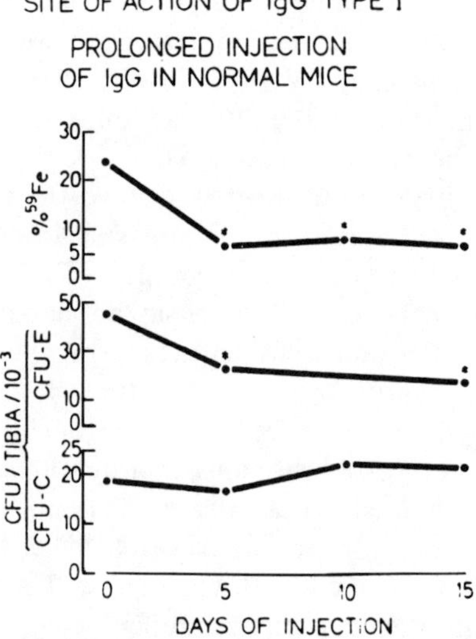

Figure 8 The site of action of IgG inhibitor of erythropoiesis in mice. Normal mice were injected with PRCA IgG for 1–2 weeks. The production of newly formed red cells was measured by the amount of ^{59}Fe incorporated into red cells, and the numbers of erythroid (CFU-E) and myeloid (CFU-C) colony-forming units were quantified among bone marrow cells obtained from the tibia of injected animals. PRCA IgG when injected into normal mice resulted in a significant decline in the production of red cells and in the number of CFU-E per tibia but had no effect on the CFU-C. These findings suggest that the target cell of this IgG inhibitor is most likely the CFU-E or a closely related erythroid cell. (From Peschle et al., 1978.)

in normal and in exhypoxic polycythemic mice. They demonstrated that after prolonged injection of PRCA IgG to normal or polycythemic animals both the ^{59}Fe incorporation into red cells and the number of CFU-E per tibia declined significantly, whereas the number of granulocytic-monocytic progenitors (CFU-GM) remained unchanged (figure 8). The same group of investigators also studied a unique patient with PRCA and an inhibitor directed against erythropoietin rather than the erythroid marrow cells. In this patient they demonstrated that the IgG inhibitor interacts with erythropoietin in solution and both IgG and erythropoietin can be precipitated by the addition of goat antihuman gamma globulin. The serum of this patient had no erythropoietin activity when tested in the polycythemic mouse assay. However, following acidification, which dissociates the antibody from the antigen, and boiling of the serum, which precipitated the IgG but not the heat and acid resistant molecule of erythropoietin, considerable erythropoietin activity was detected demonstrating that this inhibitor was interacting with circulating erythropoietin. This inhibitor disappeared from the patient's plasma following successful treatment of PRCA with immunosuppressive agents (Peschle et al., 1975*b*).

With the exception of the last study, the results of all these experiments in vivo indicated only that the plasma of patients with PRCA contains an inhibitor of murine erythropoiesis, and that this inhibitor resides in the plasma IgG fraction. The data generated from these studies, however, were far from proof that this inhibitor was an autoantibody directed against human erythroid cells. Despite the fact that this inhibitory activity disappeared from the patient's plasma during remission of PRCA, suggesting a relation of this inhibitor with the activity of the disease, the importance of this inhibitory IgG in the pathogenesis of the disease in humans remained unknown. The possibility that these across-species experiments may have detected a heterophile antibody reacting with murine erythroid cells and totally unrelated to the pathogenesis of PRCA (a situation similar to heterophile antibodies against sheep and bovine red cells during the active phase of infectious mononucleosis) remained open.

The Effect of PRCA Plasma on Heme Synthesis by Human Marrow Cells in Vitro

In 1966, Barnes reported that using a short-term culture of compatible bone marrow cells, he was able to demonstrate the presence of an inhibitor in the sera of four patients with pure red cell aplasia and thymoma. This inhibitor diminished the nuclear uptake of ^{3}H-thymidine and the cytoplasmic uptake of ^{59}Fe into erythroblasts. In two of the investigated patients he suggested that the inhibitory substance was an immunoglobulin capable of inhibiting the growth and proliferation of bone marrow cells in vitro. Details of the methods used in these studies were not reported (Barnes, 1966).

In 1963, Krantz and colleagues described a cell culture system in which human bone marrow cells responded to erythropoietin by increasing the rate of the heme synthesis (Krantz et al., 1963; Krantz, 1965). In this cell culture system, human marrow cells are suspended in tissue culture medium containing human plasma or fetal calf serum and erythropoietin and are incubated at 37° C in a 5 percent CO_2 atmosphere. The rate of heme synthesis by cells in culture is quantified by pulsing the cultures with ^{59}Fe previously bound to transferrin. At the end of the pulse period, the cells are lysed, the heme is extracted from the cell lysate, and the radioactivity is measured in the heme extract. The amount of ^{59}Fe incorporated into newly synthesized heme by erythroid cells is proportional to the number of plated cells and the concentration of erythropoietin in the medium, and is inversely proportional to the concentration of iron in the culture medium. In the presence of optimal concentrations of erythropoietin (0.25–0.5 U/ml depending on the preparation) human marrow cells after 24 to 48 hours in culture increase their rate of heme synthesis as compared to cultures without erythropoietin. This effect of erythropoietin on heme synthesis by bone marrow cells peaks after 60 to 80 hours in culture and then declines. Normal marrow cells respond to erythropoietin by a minimum of two to threefold increase in the rate of heme synthesis at 40 to 60 hours in culture. In this system the events of terminal erythroid differentiation induced by erythropoietin can be reproduced in vitro, and factors that affect the maturation of proerythroblasts to late orthochromatic erythroblasts can be examined.

Using the above-described human marrow cell culture system, Krantz and Kao (1967) studied the effect of PRCA plasma on the heme synthesis in vitro by the patient's own marrow cells and normal marrow cells. When the patient's marrow cells were cultured in vitro in the presence of erythropoietin and normal plasma they increased the rate of the heme synthesis after 72 hours threefold, showing a normal response to erythropoietin. When normal plasma was replaced by the patient's plasma, significant inhibition of the heme synthesis was noted which was proportional to the concentration of PRCA plasma present in the culture medium (figure 9). Since dilution of PRCA plasma with a salt solution allowed a normal response to erythropoietin by autologous and normal marrow cells the authors concluded that this inhibitor of heme synthesis was acting on erythroid marrow cells rather than on erythropoietin. This study demonstrated that despite the conspicuous absence of erythroblasts from the marrow of PRCA patients, their marrow cells are capable of responding to erythropoietin in vitro, but such a response is prevented by a factor present in their plasma. This was the first evidence for the presence of an autoantibody in the plasma of patients with

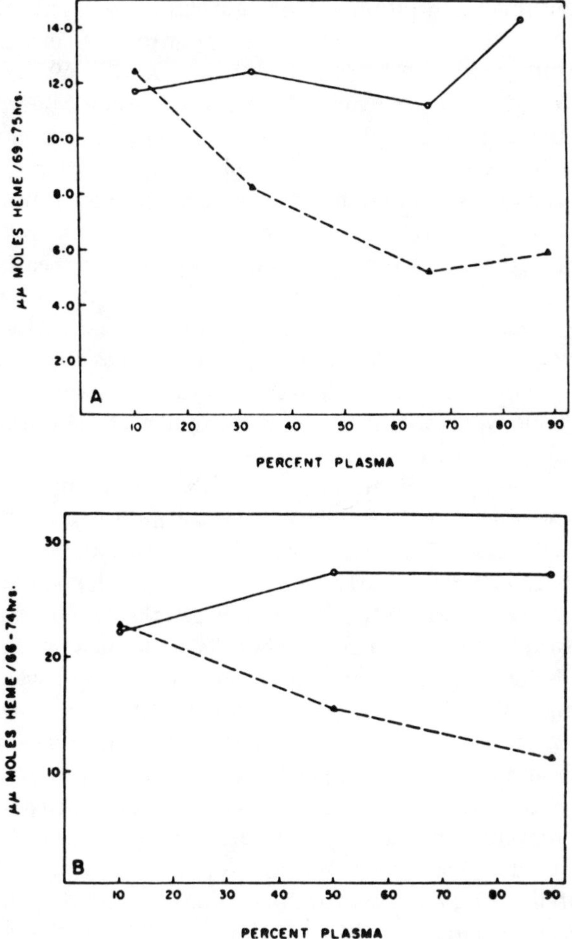

Figure 9 The effect of various dilutions of plasma from a patient with pure red cell aplasia (············) and normal plasma (————) on heme synthesis by (*A*) autologous and (*B*) normal erythroid marrow cells. Increasing the concentration of PRCA plasma in the culture medium resulted in a decrease in the amount of heme synthesized by autologous or normal marrow cells as compared with normal plasma. (From Krantz and Kao, 1967.)

PRCA acting on human autologous and normal erythroid marrow cells. A normal response of PRCA marrow cells to erythropoietin in vitro has been reported in other cases of PRCA (Krantz and Kao, 1969; Safdar et al., 1970; Krantz, 1972*a*, 1972*b*; Krantz, 1974; and Krantz and Dessypris, 1981) (figure 10). In addition, morphologic examination of the patients' marrow cells after 48 hours in liquid suspension culture in the presence of erythropoietin has shown a sevenfold increase in the number of erythroblasts, including mature orthochromatic erythroblasts absent in the initially plated cell suspension, indicating that a factor in vivo prevents the early erythroid precursors or their progenitors from maturing into late orthochromatic erythroblasts (Zaentz et al., 1975). In a group of 24 patients with PRCA whose marrow was studied in vitro by the heme synthesis system, a normal response to erythropoietin was found in 14 cases (a twofold to ninefold increase in heme synthesis rate) and a subnormal response in the remaining 10 (Krantz and Dessypris, 1981).

The presence of an inhibitor of heme sythesis by human marrow cells in vitro was demonstrated in the plasma of many more cases of PRCA. The plasma of two additional patients with PRCA was studied in vitro and found to inhibit heme synthesis by normal marrow cells and the inhibitory activity was localized to the IgG fraction. When the patient's purified IgG was added to cultures of normal marrow cells it inhibited by 75 percent the rate of the heme synthesis. This inhibitory effect of IgG was no longer present in the patient's sera after remission of PRCA (Krantz and Kao, 1969; Krantz, 1972*a*). Similar findings were reported in additional cases of PRCA. Inhibition of heme synthesis by human marrow cells in vitro by the plasma or serum IgG fraction of PRCA patients was detected in 6 out of 6 cases studied by Jepson and Vas (1974) and in 14 out of 17 Italian patients (Peschle et al., 1978) as well as in other isolated cases of PRCA (Peschle and Condorelli, 1972; Zalusky et al., 1973; Zucker et al., 1974; Marmont et al., 1975: Peschle et al., 1975*a*; Dessypris et al., 1981). It should be noted, however, that a number of these early studies were performed with normal and not autologous marrow cells, which raises questions about the significance of the findings in these cases. Since PRCA patients receive multiple red cells transfusions they may develop antibodies to various blood group antigens, present in red cells and erythroblasts, and these antibodies, included in their serum IgG fraction, may attach to normal marrow cells and in a nonspecific way inhibit heme synthesis. Therefore, such studies today cannot be considered meaningful unless they are performed on a patient's own marrow cells or exceptionally on normal cells only if the plasma, serum, or IgG tested is from blood drawn before any transfusion is given. In addition the degree of inhibition must be significant, greater than 50 percent of normal control serum of IgG, since a fluctuation of 40 to 50 percent above or below the control can occasionally be seen with normal IgG tested

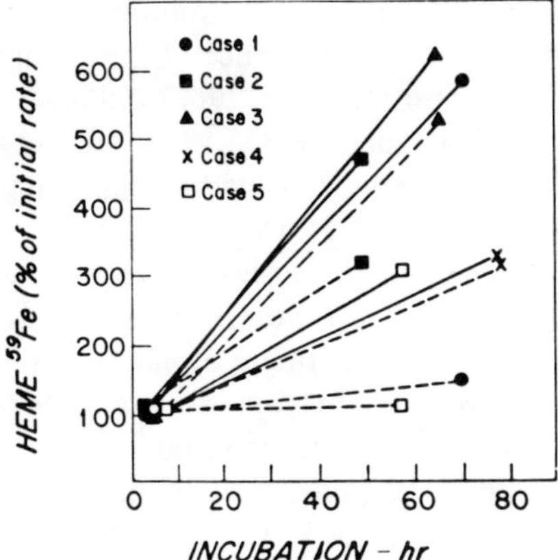

Figure 10 The response to erythropoietin (0.3–0.6 U/ml) of marrow cells from five patients with pure red cell aplasia, as assessed by the rate of ^{59}Fe-heme synthesis. Broken lines represent cultures without erythropoietin. Solid lines represent cultures containing erythropoietin. In four out of five cases, the addition of erythropoietin to cultures of PRCA bone marrow cells resulted in an increased rate of heme synthesis despite the absence of morphologically recognizable erythroblasts among the plated cells. (From Krantz, 1972*a*.)

in an autologous system (Dessypris and Krantz, unpublished observations). Following these criteria, an IgG inhibitor of heme synthesis by autologous marrow cells was detected in four out of ten patients studied at Vanderbilt University Hospital (Krantz and Dessypris, 1981). In general the degree of inhibition was proportional to the concentration of IgG in the culture medium and the inhibitory activity disappeared from the IgG fraction after remission of the disease. The incidence of IgG inhibitor in the plasmas of PRCA patients is difficult to estimate since the number of patients studied appropriately is small, but a realistic estimate should fall within the range of 30 to 40 percent.

The Effect of PRCA Plasma on Marrow Erythroid Precursors

Erythroblast Cytotoxicity Assays

Following the demonstration of an inhibitor of heme synthesis by human marrow cells in vitro in the plasmas of patients with PRCA and its further characterization as a gamma immunoglobulin (IgG), questions were raised regarding its mode of action. Previous studies in experimental animals and on human marrow cells in short-term liquid cultures have suggested that this immunoglobulin inhibitor most likely acts on erythroid precursors and does not interfere with the action of erythropoietin on marrow erythroid cells (vide ante). Using immunofluorescence techniques, earlier studies have demonstrated the presence of immunoglobulins on erythroid precursor cell membranes (Barnes, 1966) and on erythroblast nuclei (Krantz and Kao, 1967, 1969; Guerra et al., 1969; Vilan et al., 1973). Other studies using similar techniques have provided conflicting results (Jepson and Vas, 1974; Marmont et al., 1975). In isolated cases membrane-bound IgG on the rare erythroblasts present in the marrow of patients with PRCA was shown either by immunofluorescence (Björkholm et al., 1976) or by the use of electron microscopy after treatment of marrow cells with staphylococcal protein A labeled with gold (Romano et al., 1980). Morphologic evidence of injury to erythroblasts in the bone marrow of a patient with PRCA has been provided also by electron-microscopic studies. In a case studied by Böttiger and Rausing (1972), the rare erythroblasts found in the patient's marrow showed pycnotic nuclei, disruption of the cytoplasmic membrane, and clumping of cytoplasmic and nuclear structures. These cells that were thought to belong to the erythroid series were sometimes found to be engulfed by reticular cells (macrophages). Phagocytosis of erythroblasts by iron-loaded macrophages has also been reported in a case of acquired PRCA described by Marinone and co-workers (1981).

In order to study the effect of PRCA serum IgG on marrow erythroblasts, Krantz and associates have developed an erythroblast cytotoxicity assay (Krantz et al., 1973; Zaentz and Krantz, 1973; Zaentz et al., 1977). In this assay, bone marrow erythroid cells are labeled with ^{59}Fe previously bound to transferrin. Afer washing away the unincorporated ^{59}Fe, labeled marrow cells are exposed to normal or PRCA serum IgG and fresh human AB serum as a source of complement. Following incubation for 4–12 hours at 37° C the cells are centrifuged and the ^{59}Fe radioactivity in the supernatant is measured. The percentage of ^{59}Fe released in the medium from the radiolabeled marrow cells represents the ^{59}Fe release index. This index measures the magnitude of injury to radiolabeled erythroblasts by the PRCA or other IgG in the presence of complement. The release of ^{59}Fe into the medium was shown to represent a release of radioactive hemoglobin and not a simple elution of the radioisotope from marrow erythroblasts. The specificity of the cytotoxic effect of IgG on erythroblasts was demonstrated by studying its effect on ^{51}Cr-labeled red cells or lymphocytes. Using this erythroblast cytotoxicity assay Krantz and associates showed that the serum of a patient with PRCA contained an IgG capable of injuring autologous and normal marrow erythroblasts but not red cells or lymphocytes. This cytotoxic to the erythroblasts IgG disappeared progressively from the patient's serum after treatment with cyclophosphamide and prednisone (Krantz et al., 1973) (figures 11 and 12). In order to delineate further the mode of action of this cytotoxic for the erythroblasts IgG a two-stage erythroblast cytotoxicity method was developed (Zaentz and Krantz, 1973;

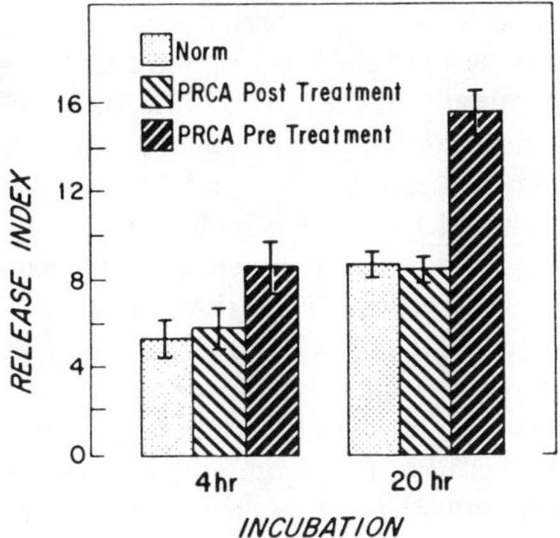

Figure 11 The effect of pretreatment and posttreatment plasma from a patient with pure red cell aplasia on the release of ^{59}Fe into the medium from radiolabeled autologous marrow erythroblasts (erythroblast cytotoxicity assay) as compared to normal plasma. (From Krantz, 1974.)

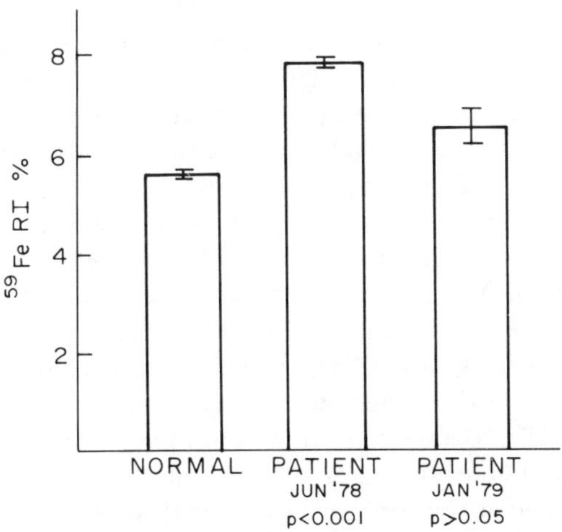

Figure 12 The effect of serum IgG from a patient with chronic granulocytic leukemia and erythroid aplasia on the release of ^{59}Fe from autologous radiolabeled erythroblasts during the phase of erythroblastopenia (June 1978) and after the appearance of erythroblasts in the patient's marrow (January 1979). (From Dessypris et al., 1981.)

Zaentz et al., 1977). In this assay ^{59}Fe-labeled erythroblasts are first coated with PRCA or normal serum IgG. The unbound IgG is washed out and then the IgG-coated cells are exposed to fresh human serum as a source of complement. In this assay inhibition of action of complement by treatment of fresh human serum by heat at 56°C for 1 hour, EGTA, suramin, or zymosan prevented the release of ^{59}Fe-hemoglobin from erythroblasts coated with PRCA IgG, indicating that the IgG-induced injury to radiolabeled erythroid precursors was mediated through the action of the complement system (Zaentz and Krantz, 1973). An improvement in the sensitivity of the erythroblast cytotoxicity assay was reported when the marrow cells were pretreated with glutathione (Zaentz et al., 1977). The above studies suggested that the cytotoxic factor present in the sera of patients with PRCA may be an antibody or an immune complex. Using this assay the effect of plasma from seven patients with PRCA on normal marrow erythroblasts was studied. Four of the seven plasmas produced an increased release of ^{59}Fe, whereas seven out of seven normal plasmas had no effect. Successive labeling of all marrow cells and erythrocytes with ^{51}Cr, a nonspecific cell label, and of marrow erythroblasts with ^{59}Fe showed that PRCA serum IgG was capable of increasing the release of ^{59}Fe but not of ^{51}Cr, a finding suggesting that this antibody or immune complex injures preferentially the erythroblasts and has no effect on erythrocytes or other nonerythroid cells in the marrow. In these studies the role of multiple red cell transfusions in the production of cytotoxic antibodies present in the plasmas of patients with PRCA was addressed by studying the ^{59}Fe release from normal erythroblasts coated with the plasma of seven multiply transfused patients without PRCA who had received at least two transfusions over the preceding 12 months. Only the plasma of one of these polytransfused patients increased the ^{59}Fe release index and this patient was rejecting a transplanted kidney at the time of the study.

Further studies on the sensitivity and specificity of the erythroblast cytotoxicity assay performed with normal ^{59}Fe-labeled marrow erythroblasts have shown that this assay is capable of detecting antierythroblast antibodies but not necessarily autoantibodies that are pathogenetically related to the disease. This became clear when plasmas cytotoxic for normal marrow erythroblasts were tested on autologous marrow erythroblasts after remission of PRCA and were found to have no toxic effect on the patient's own erythroid precursors. Thus the results of this assay can be considered of pathogenetic significance only when this test is positive with the patient's own marrow cells (Krantz and Dessypris, 1981). Using a completely autologous system an erythroblast cytotoxic IgG autoantibody or immune complex has been detected in two patients with erythroid aplasia, in one with primary acquired PRCA (Krantz et al., 1973), and in one patient with chronic granulocytic leukemia and erythroid aplasia (Dessypris et al., 1981). Using the same autologous system the sera of 31 patients with myelodysplastic syndromes (16 with refractory anemia without excess of blasts, 11 with primary acquired sideroblastic anemia, and 4 with refractory anemia with excess of blasts) who had received multiple transfusions over the preceding years were tested for the presence of autoantibodies cytotoxic to erythroblasts. In none of them was such an antibody detected, indicating that the erythroblast cytotoxicity assay, when positive on the patient's own marrow cells, may detect an autoantibody that is not related to red cell transfusions and which may be of pathogenetic significance in PRCA (Dessypris and Krantz, 1985*b*; unpublished observations). The frequency of such antibodies in the sera of patients with PRCA is not exactly known, but it does not seem to be high, since in only 2 out of 13 patients with erythroid aplasia could such an autoantibody be detected (Krantz et al., 1973; Dessypris et al., 1981; Krantz and Dessypris, 1981).

The erythroblast cytotoxicity assay can detect antibodies directed against marrow erythroblasts that can cause damage to these cells through the action of complement. Complement-mediated immune injury is not the only mechanism through which the immune system exerts its action. Target cells coated by an antibody can be injured by a cytotoxic mechanism that involves lymphocytes but does not require the presence of complement (Perlmann et al., 1972; Wunderlich et al., 1975). This is known as antibody-dependent cellular cytotoxicity (ADDC), or lymphocyte-dependent antibody cytotoxicity, and is one of the most efficient mechanisms for immune injury of target cells requiring extremely low concentrations of antibody. Such a method for demonstration of cytotoxic antibody to normal marrow erythroblasts has been developed and applied to the study of antibody-mediated injury in PRCA. In this method normal marrow cells are enriched three to fourfold in erythroblasts which are then labeled with ^{59}Fe. The radiolabeled erythroblasts are coated with plasma or IgG from patients with PRCA or normal donors and mixed with autologous blood lymphocytes (effector cells) in a ratio of 1:60. After four hours of incubation at 37° C the release of ^{59}Fe in the supernatant indicates a toxic effect of lymphocytes on IgG-coated radiolabeled erythroblasts. PRCA plasma or IgG as compared to normal controls caused a five to tenfold increase in the release of ^{59}Fe in the supernatant. This increase in the ^{59}Fe release was complement-independent but absolutely dependent on the presence of lymphocytes. PRCA plasmas or IgG were found to have such an activity against normal erythroblasts. However, the lymphocyte-mediated erythroblast cytotoxicity in the presence of IgG was not related to the activity of the disease, since it was found to be present in the plasma or IgG prepared from blood drawn from the same patients in remission. More important, when these studies were performed on autologous marrow erythroblasts they were negative, indicating that the detected antibody-dependent cytotoxicity mediated through lymphocytes was due to the presence of transfusion-acquired alloantibodies and did not appear to be pathogenetically related to the disease (Krantz and Dessypris, 1982). The negative findings of these studies do not distinguish between absence of such an antibody activity in PRCA plasmas and/or absence of any role of the selected type of effector or target cells. It is conceivable that monocytes-macrophages and not lymphocytes may play the role of effector cells in the PRCA marrow since phagocytosis of proerythroblasts in PRCA marrows has been reported in at least three cases of this syndrome (Böttiger and Rausing, 1972; Croles and van Delden, 1975; Marinone et al., 1981). It is also possible that an erythroid cell earlier than the proerythroblast is the target cell for such cytotoxic antibodies rather than the mature erythroblasts themselves, which may also explain the absence of erythroblasts in the marrow of patients with PRCA in whom a normal to increased number of the erythroid progenitor cells, CFU-E (colony-forming unit-erythroid) and BFU-E (burst-forming unit-erythroid), are assayed in vitro.

Erythroid Progenitors

During the last two decades the use of semisolid media for cloning and quantification of erythroid progenitors has provided significant information regarding the regulation of normal erythropoiesis and has allowed a new insight into disorders of red cell production. Using these methods erythroid progenitors in the marrow, morphologically unrecognizable by routine polychrome stains, can be assayed and quantified on the basis of their ability to give rise to colonies of mature recognizable erythroblasts. In human marrow at least two types of erythroid progenitors have been recognized, the CFU-E and the BFU-E.

The CFU-E is a cell close to proerythroblast which, in the presence of relatively small concentrations of erythropoietin, is capable of proliferating and differentiating into small colonies of 8–64 mature, well-hemoglobinized erythroblasts within a period of 6–8 days in culture. The BFU-E is an early erythroid progenitor, closer to the pluripotent hematopoietic stem cell which, in the presence of high concentrations of erythropoietin and other hematopoietic factors, proliferates and differentiates into large colonies of a hundred to many thousands of mature erythroblasts within 14 to 16 days in culture. A schema of erythropoiesis as currently understood is presented in figure 13, and the reader is referred to recent reviews on this subject for details (Peschle, 1980; Eaves and Eaves, 1985). The application of these methods for assaying erythroid progenitor cell growth and maturation in vitro has provided new and valuable information regarding the stage at which the arrest of erythropoiesis occurs in PRCA and has allowed a better definition of the target cells for the ery-

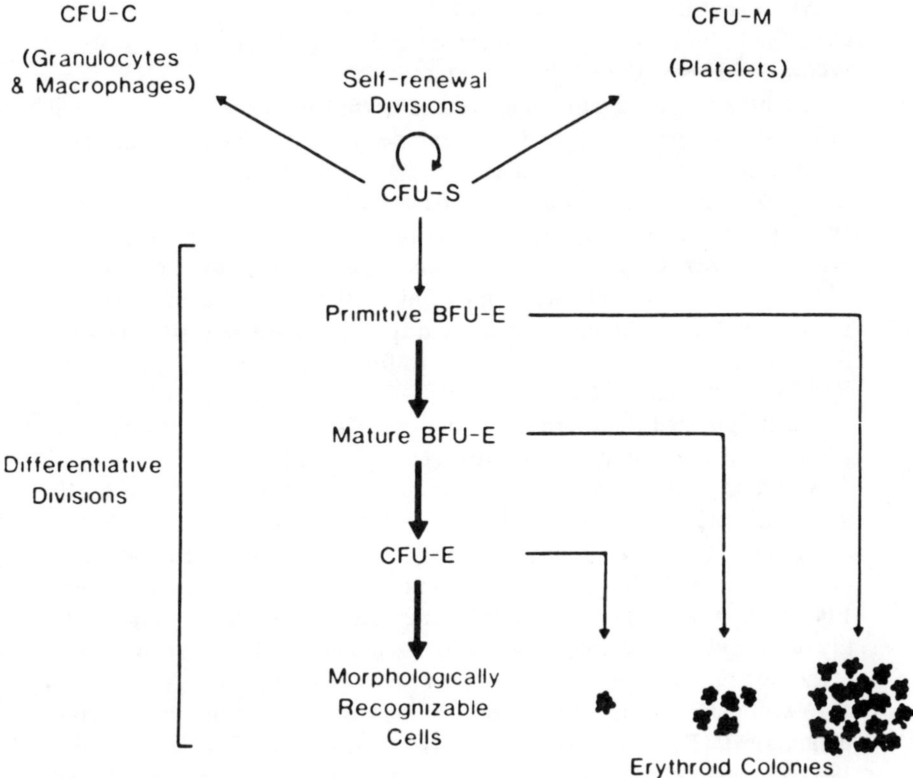

Figure 13 A schematic representation of the various steps in erythroid cell differentiation, from the pluripotent stem cell (CFU-S) to morphologically recognizable erythroblasts. (From Eaves and Eaves, 1985.)

thropoietic inhibitor present in these patients' plasmas.

The proliferative capacity of erythroid progenitors CFU-E in the marrow of twenty-one patients with PRCA was examined by Katz and coworkers (1981). This group of patients was highly heterogeneous and included five adult patients with acquired idiopathic PRCA, seven children (three with transient erythroblastopenia of childhood and four with Diamond-Blackfan anemia), one patient with aplastic crisis in the course of sickle cell anemia, two patients with drug-induced PRCA, three patients with erythroid aplasia associated with chronic lymphocytic leukemia, and two patients with erythroid aplasia that terminated in acute leukemia. In eleven of these cases, the patient's bone marrow cells produced a normal or increased number of erythroid colonies in vitro as compared to normal adult bone marrow cells, whereas in the remaining ten patients, erythroid colony formation by their bone marrow cells was either very low or absent (Katz et al., 1981). This study indicated that in a number of cases of isolated erythroid aplasia, the pool of marrow erythroid progenitors CFU-E remains intact suggesting that the arrest of erythropoiesis occurs at a level of differentiation between the CFU-E and the morphologically recognizable erythroblast. In the remaining cases of PRCA the CFU-E pool is significantly diminished, indicating that the arrest of erythropoiesis occurs at a level of an erythroid progenitor cell more immature than the CFU-E.

In another study by Lacombe and colleagues (1984), the marrow cells of twenty-two adult patients with chronic PRCA, two patients with thymoma, and three patients with erythroid aplasia associated with myeloproliferative disorders were assayed in vitro for CFU-E and BFU-E. On the basis of the number of CFU-E and BFU-E grown in vitro these investigators distinguished three groups of patients (I-III). Group I consisted of patients with PRCA whose marrow cells produced a normal number of CFU-E and a normal or mildly decreased number of BFU-E. Group II included patients with low but detectable numbers of CFU-E and BFU-E, and group III patients qualified by having undetectable CFU-E and BFU-E in their marrows. Almost 50 to 60 percent of patients with primary chronic PRCA, as well as the two studied patients with PRCA associated with thymoma, were classified into group I, again indicating that at least in half of the patients with this syndrome the arrest of erythropoiesis seems to occur at a level between the CFU-E and proerythroblast. Among twelve adult patients with PRCA referred to Vanderbilt University Hospital between 1978 and 1985, five were found to have normal numbers of CFU-E and BFU-E growth from their marrow cells in vitro, two had normal numbers of BFU-E but low or undetectable CFU-E, and the remaining five had a very low or undetectable number of CFU-E and BFU-E in their marrow cells assayed by the plasma clot method (Dessypris and Krantz, unpublished observations). It should be noted that despite the variability in the level at which erythropoiesis is arrested, clinically the disease looks the same and no clinical differences can be appreciated among patients whose erythropoiesis is arrested at an early stage versus those whose erythropoiesis is arrested at a later stage.

It has been suggested that the response of the patient's marrow cells to erythropoietin in vitro as assessed by the heme synthesis method, or the presence in the patient's marrow of a normal number of erythroid progenitors, is associated with a higher response rate to immunosuppressive therapy compared to the response rate of patients whose marrow does not respond in vitro to erythropoietin or lacks in vitro assayable erythroid progenitors (Krantz, 1974; Lacombe et al., 1984). Although the above observations are correct when applied to the whole population of patients with PRCA, we feel that simple quantification of CFU-E and BFU-E in the marrow of patients with PRCA can be used to identify the level at which arrest of erythropoiesis occurs, but it cannot be used to predict the presence or absence of a serum IgG inhibitor or the response of an individual patient to immunosuppressive treatment. Three of our patients with undetectable erythroid progenitors in their marrow responded to immunosuppressive treatment. In two of them the number of erythroid progenitors in the marrow returned to normal after remission of their disease. The third patient continued to have very low numbers of erythroid progenitors in his marrow despite a normal hematocrit. This patient developed acute myelomonocytic leukemia six months after recovery from the PRCA. The presence of undetectable erythroid progenitors in the marrow of a number of patients with PRCA cannot be interpreted as evidence for a "stem cell defect" or as evidence against an immune pathogenesis since arrest of erythropoiesis may occur at an earlier level resulting in severe depletion of the erythroid progenitor pools in the marrow. Alternatively in those cases in which humoral inhibitors of erythropoiesis can be detected after recovery of PRCA it seems that the low or undetectable erythroid progenitor cell growth in vitro during the active phase of the disease may be the result of injury of these erythroid cells occurring in vivo that results in loss of their ability to proliferate and differentiate into erythroblastic colonies in vitro.

In general the arrest of erythropoiesis in PRCA seems to occur at any level between the early BFU-E and the early basophilic erythroblast. Quantification of CFU-E and BFU-E numbers in the marrow of patients with PRCA indicates that in at least 60 percent of cases the CFU-E and BFU-E pools remain intact, suggesting that in these cases the arrest of erythroid differentiation in vivo occurs at the level of the CFU-E, or between the CFU-E and the proerythroblasts. In the remaining cases of erythroid aplasia the erythroid progenitor cell compartment is affected at a stage earlier than the CFU-E stage and the CFU-E and/or BFU-E marrow pools are significantly reduced. As a rule, following recovery of erythropoiesis and restoration of the red cell mass to normal levels the number of CFU-E and BFU-E in the marrow returns to normal (Lowenberg and Ghio, 1977; Dessypris et al., 1984).

The Effect of PRCA Serum or Serum IgG on Erythroid Progenitors

The availability of clonal assays that allow the study of development of recognizable erythroblasts from their morphologically indistinguishable progenitors, CFU-E and BFU-E, has generated a number of studies in which the effect of plasma, serum, or serum IgG from patients with PRCA on erythroid colony development in vitro was examined. Browman et al. (1976) reported first that the serum from a patient with PRCA and Hashimoto thyroiditis was capable not only of inhibiting erythropoietin-induced heme synthesis by normal marrow cells but also of suppressing erythroid colony formation from normal and autologous CFU-E in vitro in a system depleted of any significant complement activity. They were unable, however, to document any relation of in vitro findings with the activity of the disease. In another study of a patient with PRCA and systemic lupus erythematosus, Cavalcant and co-workers (1978) showed that the patient's serum had a markedly inhibitory activity on autologous erythroid colony formation in vitro. Katz and colleagues (1981) also found that serum from four out of five patients with acquired PRCA when added to the culture medium inhibited the growth of autologous bone marrow CFU-E by 65 to 100 percent. In another case of PRCA, Messner and associates (1981) showed that the patient's plasma contained an inhibitor of autologous and normal blood BFU-E growth in vitro. This inhibitory activity was localized to the IgG fraction, was specific for erythroid and not granulocytic/monocytic colony formation (CFU-GM), and was related to the disease activity, since IgG purified from plasma collected after plasmapheresis-induced remission of PRCA had no demonstrable effect on erythroid colony formation in vitro (figure 14). In their experiments, they also demonstrated that the decrease in the number of blood BFU-E–derived colonies was most likely due to an effect of IgG on a cell more mature than the BFU-E, since addition of IgG to the culture medium as late as day 6 to 7 of culture, a period during which BFU-E have matured to CFU-E, resulted in a decline of erythroid colonies equal in magnitude to that seen when the IgG was incorporated in the medium at the initiation of the cultures. Lowenberg and Ghio (1977) attempted to characterize the site of action of this inhibitory IgG by incubating marrow cells with normal or PRCA serum with or without complement and then assaying them for CFU-E. They concluded that inhibition of erythropoiesis in PRCA is achieved by a complement-dependent plasma factor which eliminates or inactivates CFU-E. These experiments, however, were performed on normal marrow cells and the inhibitory factors may have been transfusion-acquired antibodies. In addition, the preincubation step with the patient's plasma, not followed by removal of the plasma by washing the marrow cells, does not eliminate the possibility that the IgG acts on a CFU-E–derived cell and not the CFU-E per se. In a group of nine children with red cell aplasia (TEC) a serum inhibitor of autologous CFU-E and/or BFU-E growth was detected in four out of six evaluable cases; in the remaining three, CFU-E growth was so poor that no conclusion could be drawn regarding the presence of an inhibitor. Purified serum IgG was found to be inhibitory for autologous CFU-E and BFU-E growth in three cases and for normal marrow CFU-E and BFU-E in four and six cases, respectively. A relation of inhibitory activity with the active phase of the disease and demonstration of a direct cytotoxic effect on CFU-E per se were also shown (Dessypris et al., 1982). It should

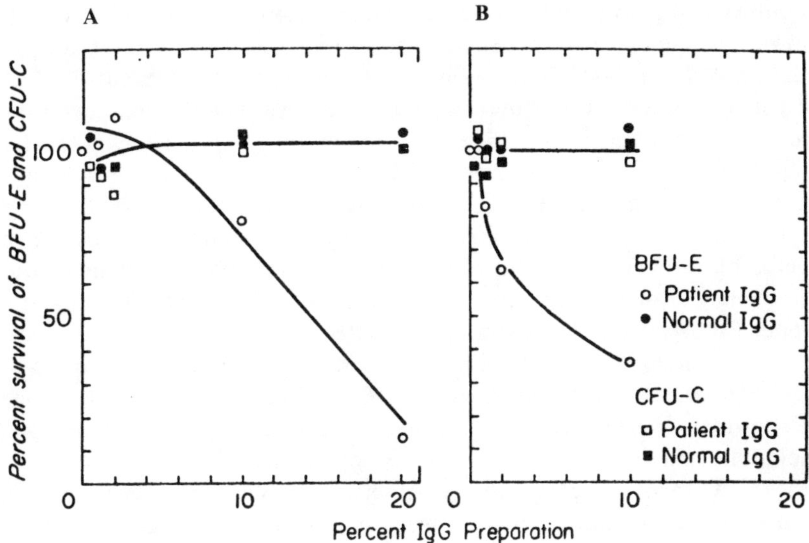

Figure 14 Erythroid burst (BFU-E) and granulocytic (CFU-C) colony formation by (*A*) autologous and (*B*) normal bone marrow cells in the presence of pure red cell aplasia and normal serum IgG. (From Messner et al., 1981.)

be mentioned that this was the first demonstration of a direct cytotoxic effect on autologous CFU-E per se since all previous studies with cultures of erythroid progenitors could have observed effects on the late cells which would have resulted in counting reduced numbers of erythroid colonies, attributed to decreased CFU-E numbers. Among nine adult patients with acquired PRCA who were studied at Vanderbilt during the period 1980 to 1984, a serum IgG inhibitory for autologous marrow CFU-E and/or BFU-E growth in vitro was detected in six. In three of these six patients, preincubation of autologous marrow cells with IgG and complement followed by removal of the IgG and complement by repeated washing of marrow cells resulted in a significant (greater than 50 percent) decline in the numbers of CFU-E grown in vitro compared to the number of CFU-E grown from autologous marrow cells preincubated with complement alone or normal IgG with or without complement (Dessypris et al., 1984, 1985; Dessypris and Krantz, unpublished observations). It seems that in a number of patients with acquired PRCA inhibition of erythroid colony formation by the patient's own marrow cells in vitro can be demonstrated in the presence of the patient's IgG. In some cases this IgG has cytotoxic activity against the CFU-E per se and inhibition of CFU-E growth is seen only in the presence but not in the absence of complement. In other cases inhibition of growth and maturation of CFU-E or of a CFU-E–derived cell in vitro is complement-independent and requires the presence of IgG during the whole 7 to 8 day period of cultures. The absence of an inhibitory effect on heme synthesis by marrow cells and the absence of erythroblast cytotoxicity indicates that in such cases the IgG inhibitor most likely exerts its effect on the CFU-E per se or on a CFU-E–derived more mature erythroid progenitor but not on mature erythroblasts. Whether the target cell for the inhibitor remains the same during the various phases of the disease is not yet known. Since the number of patients studied thus far is very small, the frequency of such an inhibitor in the serum of patients with acquired primary PRCA cannot be accurately estimated.

The interpretation of findings from in vitro cloning of marrow erythroid progenitors in the presence of PRCA serum or IgG requires caution. Experiments performed on normal marrow cells and not confirmed on autologous marrow cells cannot be considered conclusive. Furthermore in these in vitro assays IgG purified from serum of single normal donors has a variable effect on autologous erythroid colony formation ranging from a mild inhibition of 35 percent to a stimulatory effect as significant as 25 percent as compared to cultures containing medium alone. Therefore comparison of the effect of PRCA IgG should preferably be done with IgG purified from pooled normal serum of 20 or more donors. In addition, taking into account the variability of the effect of normal IgG on autologous marrow erythroid colony formation in vitro, an inhibitory effect cannot be considered definitely significant unless it reaches the range of 50 percent.

Lymphocyte-mediated Suppression of Erythropoiesis

Despite extensive attempts to document a humoral inhibitor of erythropoiesis in the serum of patients with PRCA, there remains a substantial proportion of cases in which such an inhibitor cannot be demonstrated by anyone using currently available laboratory techniques. In those cases suppression of erythropoiesis by lymphocytes has been thought to be the probable mechanism for the production of erythroid aplasia. In 1977, Litwin and Zanjani reported the presence in the peripheral blood of two patients with thymoma and hypogammaglobulinemia of lymphocytes that suppressed in vitro both immunoglobulin production and erythroid differentiation. As they pointed out, however, the presence of these lymphocytes did not correlate with the course of erythroid aplasia; one of their patients possessed such suppressor cells for erythroid development while in remission, and the other patient had cells with similar activity although he was hematologically normal and had never developed red cell aplasia. They concluded that it is unlikely that these lymphocytes were the sole cause of anemia in the first patient and they emphasized that caution must be exercised in interpreting in vitro data indicating that these cells play a primary etiologic role in the pathogenesis of this disorder.

A case of erythroid aplasia that developed in the course of T cell chronic lymphocytic leukemia has been investigated by Hoffman and coworkers (1978). In this case the patient's peripheral blood or bone marrow T-lymphocytes suppressed normal marrow CFU-E growth in vitro by 70 percent. The ratio of T-lymphocytes added to normal marrow cells was almost 1:1, corresponding to the ratio of T-lymphocytes to all other cells in the patient's bone marrow, which was infiltrated by T-lymphocytes comprising 56 percent of the marrow-nucleated cells. This suppressive effect on erythroid-colony formation in vitro was not seen when an equal number of normal lymphocytes or lymphocytes from a B cell chronic lymphocytic leukemia were added to marrow cells. In addition the patient's own marrow cells were assayed in vitro for their erythroid colony–forming capacity. No growth of patient's CFU-E was detected using whole marrow cells, but after treating the marrow cell suspension with antithymocyte globulin and complement, which lysed the T-lymphocytes, the growth of CFU-E increased to $84 \pm 17/6 \times 10^5$ plated cells. These studies were performed before any red cell transfusions were given to the patient so the suppressive effect of T-lymphocytes could not be attributed to transfusion-generated suppressor/toxic cells (Hoffman et al., 1978). These findings were confirmed and extended by Nagasawa et al. (1981), who studied a case of T cell chronic lymphocytic leukemia in which the T-lymphocytes had membrane receptors for the Fc portion of the IgG molecule (T_γ). In this case the patient had developed both erythroid aplasia and hypogammaglobulinemia during the course of his leukemia. Malignant T_γ cells suppressed erythroid colony formation in vitro in a dose-dependent fashion but had no effect on the growth of CFU-GM in vitro. Their suppressive activity was demonstrable both on normal and autologous bone marrow cells as well as on normal and autologous B-lymphocytes. These experiments also suggested that the inhibition of erythroid colony formation in vitro requires direct cell to cell contact (interaction) between erythroid progenitors and T-lymphocytes, since medium conditioned by T cells stimulated by phytohemagglutinin (PHA) had no activity at all. This was an important piece of information suggesting that the findings from experiments on the suppressive role of lymphocytes on erythroid colony formation in vitro may have in vivo significance only when one can demonstrate that an appropriate proportion of suppressive T-lymphocytes is present in the patient's bone marrow. Furthermore, the effects of peripheral blood T cells on erythroid colony formation in vitro cannot be pathophysiologically meaningful unless the patient's marrow is infiltrated by lymphocytes with the same surface markers as blood lymphocytes, or unless it can be demonstrated that their effect on erythroid progenitors is mediated through a humoral factor released from these lymphocytes.

A number of additional cases of erythroid aplasia have been described in which a suppressive effect of blood or marrow T-lymphocytes has been demonstrated in vitro and a pathogenetic role for the induction of erythroid aplasia has been suggested. In a case described by Linch et al. (1981) the patient's differential white blood count showed a lymphocytosis with 76 percent lymphocytes, and his bone marrow was infiltrated by lymphocytes constituting 45 percent of the nucleated marrow cells. Therefore this case should be classified as erythroid aplasia secondary to T cell chronic lymphocytic leukemia or, as some authors prefer to call it, chronic T cell lymphocytosis. In this case depletion of marrow cells of T-lymphocytes resulted in a significant increase of the number of CFU-E far beyond what one would expect by concentrating the erythroid progenitors by the T cell depletion. Addition of T cells to T cell–depleted marrow cells resulted in suppression of erythroid colony formation by CFU-E and BFU-E; however, this suppression was not dose-dependent, and when the number of T cells to marrow cells reached a 1:1 ratio, as actually was the ratio in the patient's marrow, the inhibitory effect disappeared. In addition the presence in, and development from, whole marrow cells of normal numbers of CFU-E indicated that the simple presence of T cells may not have been solely responsible for the erythroid aplasia (Linch et al., 1981). In two additional cases of PRCA described by Hanada et al. (1984), in which a population of peripheral blood T cells was shown to exert an inhibitory effect on autologous CFU-E growth in vitro, the presence or absence of marrow lymphocytosis is not reported. Therefore, whether these patients had primary acquired PRCA or erythroid aplasia secondary to chronic T cell lymphocytosis cannot be determined. If indeed these patients had primary acquired PRCA, the significance of the suppression of erythroid colony development in vitro by blood lymphocytes remains unclear in view of the fact that such a suppression by increased numbers of T-lymphocytes requires cell to cell interaction (Nagasawa et al., 1981) which would be difficult to conceive in the absence of bone marrow lymphocytosis unless one could demonstrate that the blood T-lymphocytes release a factor inhibiting erythroid cell growth and/or maturation in vitro. The presence of an increased number of T cells in the blood of these patients may be an epiphenomenon, since such changes in the peripheral blood lymphocytes have been described in a variety of conditions. In addition, a decrease of these suppressor lymphocytes following immunosuppressive treatment and remission of PRCA cannot be pathogenetically linked to the erythroid aplasia since lymphopenia can be the result of a variety of immunosuppressive regimens.

Suppression of erythropoiesis by T cells was suggested also by the studies of Abkowitz et al. (1986). In this unusual case of PRCA the patient presented with hepatosplenomegaly. The patient's peripheral blood contained an increased number of large granular lymphocytes (35 percent of lymphocytes) and the marrow was hypocellular and was infiltrated by lymphocytes constituting 14 percent of the marrow-nucleated cells. These T cells were phenotypically characterized as immature, nonfunctional, natural killer (NK) cells. The patient's marrow cells grew no CFU-E–derived colonies, whereas the number of BFU-E and CFU-GM grown in vitro was normal. Removal of T cells from the marrow resulted in a disproportionate increase of BFU-E in vitro; however, the effect of T cell depletion on CFU-E growth was not reported. Addition of peripheral blood T cells collected during the active phase of PRCA, but not during remission, to autologous marrow cell cultures resulted in suppression of erythroid but not myeloid colony formation in vitro. The inhibition of erythroid colony formation was seen when marrow cells were cocultured with blood T cells at a ratio of 1:1. Whether inhibition could be demonstrated also at a ratio of 1:4 or 1:5, corresponding to the ratio of lymphocytes to nucleated cells in the patient's marrow, was not reported. Interestingly, conditioned medium from cultures of unstimulated or PHA-stimulated lymphocytes had no effect on erythroid burst formation in vitro suggesting that the demonstrable inhibitory effect of T cells on BFU-E growth and maturation was probably mediated through a direct cell to cell interaction. The authors concluded that in this case arrest of erythropoiesis in vivo occurred between BFU-E and CFU-E, and that this arrest was mediated by T-lymphocytes. As they pointed out, however, whether this case represented a truly primary PRCA or was secondary to a lymphoproliferative T cell disorder cannot be determined. In another case of erythroid aplasia in a patient with rheumatoid arthritis the suggested suppression of erythropoiesis by peripheral blood lymphocytes as indicated by a 90 percent increase in the number of marrow CFU-E grown in vitro after T cell depletion cannot be considered significant, since in normal control marrow a 70 percent increase of CFU-E growth was also noted following removal of normal T cells (Konwalinka et al., 1983). In an additional case of juvenile rheumatoid arthritis and pure red cell aplasia studied by Kimura et al. (1983), depletion of marrow cells of T-lymphocytes resulted in a significant increase in the number of CFU-E and BFU-E growth in vitro; however, experiments in which the purified marrow T cells were added back to T cell–depleted marrow cells were not reported. These T-lymphocytes had no effect on normal marrow erythroid colony formation in vitro.

Further insight into the mechanism of T cell–mediated suppression of erythropoiesis in vitro in T cell lymphoproliferative disorders (T cell CLL) was provided by Lipton and associates (1983). In a case of erythroid aplasia associated with T cell lymphocytosis they studied the effect of lymphocytes on autologous erythroid marrow colony formation in vitro. They attempted to further characterize the suppressor T cells by a battery of monoclonal antibodies and determine their activity as natural killer or cytotoxic lymphocytes. In this case, the suppressor cells had membrane antigens reacting with T3, T8, T11, Ia, and Mol monoclonal antibodies but not with T1 or Mo2. In addition they showed that these cells had no natural killer or cytotoxic cell activity. On Wright-Giemsa staining these cells had mixed features of a monocyte/lymphocyte. They were larger than usual lymphocytes with abundant cytoplasm containing small numbers of slightly azurophilic granules, a picture collectively described as that of a large granular lymphocyte. The suppressive effect of these cells on erythroid colony formation in vitro was genetically restricted to cases in which erythroid progenitors shared antigens at the HLA-A or DR locus. Thus, suppression of erythropoiesis could be demonstrated on autologous marrow cells and on normal marrow cells only from donors that had common HLA-A and/or DR antigens but not on those who were completely HLA histoincompatible. Moreover, the magnitude of suppression correlated with the degree of DR homology and it was more pronounced at the CFU-E than the BFU-E level. It should be noted, however, that these studies were performed with peripheral blood and not bone marrow lymphocytes and that the patient's marrow was reported as showing only occasional cells that were of indeterminate type—possibly large granular lymphocytes—on biopsy. Since the suppression of erythropoiesis was genetically restricted, the assumption that a direct cell to cell contact that leads to suppression of erythroid progenitor cell development must take place in vivo is unavoidable. However, this patient's marrow showed no or minimal lymphocytic infiltration and the ratio of lymphocytes to marrow cells required for demonstration of significant inhibition of erythropoiesis in vitro was at a minimum of 1:1, implying that at least a 50 percent replacement of the marrow must be present on biopsy to consider the in vitro experiments pathophysiologically meaningful. It should be mentioned also that the whole unseparated marrow cell suspension gave rise to a normal number of CFU-E and BFU-E, a finding suggesting that if T cells in certain cases do play a role in the pathogenesis of erythroid aplasia, their role may not be of primary significance. This view is supported by the findings of Hocking and co-workers (1983), who demonstrated that in a case of T cell CLL with 20 percent infiltration of bone marrow by lymphocytes, erythroblasts were present in the marrow and were morphologically megaloblastic. In this case effective erythropoiesis was present to the point of maintaining a red cell volume of 30 percent despite the fact that removal of T cells from the peripheral blood allowed a small number of blood BFU-E to develop into erythroblasts, a finding interpreted by the investigators as suggestive of T cell–mediated inhibition of erythropoiesis. In addition, among twenty cases of chronic T cell lymphocytosis with marrow involvement in which the lymphocytes were T3+, T8+, Fc_{γ}+, T1−, Dr + (Ia) only two had erythroid aplasia, indicating that the presence of such T cells in the marrow is not by itself sufficient to lead to erythroid aplasia (Newland et al., 1984). It also has been shown that normal individuals have among their peripheral lymphocytes subsets of these cells with opposing effects on erythropoiesis in vitro (Torok-Storb et al., 1981) and that the T cell enhancement or suppression of in vitro erythropoiesis is restricted by the HLA-DR locus (Torok-Storb and Hansen, 1982). However, these findings in vitro are applicable to cultures of peripheral blood cells, a tissue in which, under normal conditions, proliferation and development of erythroid progenitors into erythroblasts does not regularly take place as indicated by the absence of mature erythroid progenitors (CFU-E) from the null-cell fraction of peripheral blood mononuclear cells. Therefore, interactions in vitro between peripheral blood lymphocytes, of a type not present in the marrow, and erythroid progenitor cells must be interpreted with caution—especially in regard to their regulatory role in normal individuals and their pathophysiologic significance in disease states. The relatively high coexistence of erythroid aplasia and T cell CLL, both of which are rare diseases, suggests that there must be a relation between a T cell population and the erythroid development in vivo. Further investigations are necessary, however, to elucidate the factors other than the presence of T cells that are responsible for the production of erythroid aplasia.

Erythroid aplasia has not only been reported in association with T cell CLL but more frequently with B cell CLL. Actually, B cell CLL is the most common lymphoproliferative disorder with which PRCA has been associated. It is well known that PRCA associated with CLL is notorious for the absence of humoral inhibitors of erythropoiesis in the sera of these patients (Peschle et al., 1978; Krantz, SB, personal communication, 1980; Mangan et al., 1981). The pathogenesis of erythroid aplasia associated with CLL has been extensively investigated by Mangan and co-workers (1981, 1982). The sera and IgG fractions from two such patients had no inhibitory activity on autologous or normal marrow CFU-E growth in vitro. An increased number of T-lymphocytes was found in a surprisingly increased percentage (70 percent and 25 percent) in the marrow of these patients. Further investigation into the type of T cells showed that 35 to 90 percent of these T-lymphocytes expressed on their membranes receptors for the Fc fragment of IgG (T_{γ} cells). In three other patients with B cell CLL and no erythroid aplasia T_{γ} cells were found also to be increased in the marrow but their total number was much lower than in patients with PRCA. Bone marrow CFU-E from patients with erythroid aplasia associated with CLL were severely depressed, but after removal of T cells a ten-fold increase in the number of CFU-E grown in vitro was noted, an increase much higher than one would expect from simple concentration of erythroid progenitors resulting from T cell depletion. In addition coculture of the T cells with T cell–depleted autologous marrow cells at 1:1 ratio resulted in 80 to 90 percent suppression of CFU-E growth in vitro, whereas coculture of T_γ cells from normal controls had no significant effect on the CFU-E growth in vitro. T_γ cells from these patients were also found to suppress normal marrow CFU-E but not CFU-GM growth in vitro. The percentage of T_γ cells in the CLL marrow decreased significantly, returning to normal levels after chemotherapy-induced remission of PRCA and coculture experiments showed that their inhibitory activity on erythroid colony formation was no longer demonstrable in vitro. At least one of the two patients was studied before any red cell transfusion was given, excluding the possibility that the T cell suppression of erythropoiesis was induced by transfusion-mediated expansion of T-lymphocytes. In patients with untreated B cell CLL these investigators found an increased number of T_γ cells in their blood but not in their marrow. In the two studied patients with CLL and red cell aplasia an abnormal "migration" of T_γ cells in the marrow was noted. This finding provided some explanation for the fact that although T_γ cells may be increased in the blood of patients with CLL, erythroid aplasia occurs rarely, presumably only in those cases in which massive migration (and/or proliferation, accumulation) of these cells to the marrow takes place. This finding lends further support to our view that in vitro data based on coculture experiments of blood lymphocytes with marrow cells in which T cell–mediated suppression of erythropoiesis is demonstrated are of doubtful pathophysiologic significance unless infiltration of the patient's marrow by these lymphocytes can be documented. In a more recent study by Mangan and D'Alessandro (1985) it was shown that among 30 patients with B cell CLL there was a gradual accumulation of T_γ - lymphocytes with a suppressive effect on erythropoiesis in vitro, which, however, became significant only in cases with coexisting red cell aplasia. Expansion of T_γ cell population in the marrow was detectable even at the earliest stages of the disease. It progressed with evolution of the leukemia from Rai stage 0 to stage IV, and it was independent of red cell transfusions. These findings, similar to those reported earlier by Catovsky and colleagues (1981), suggest that the hypoproliferative anemia in CLL (Berlin et al., 1954; Wasi and Block, 1961) is basically related to the progressively increasing numbers of T_γ-lymphocytes accumulating in the marrow which in a number of patients reach such a high proportion that complete erythropoietic arrest occurs. The mechanism through which T_γ cells suppress erythroid progenitor growth and development remains unclear.

In addition to chronic B- or T-lymphocytic leukemia, lymphocyte-mediated suppression of erythropoiesis has been proposed as the pathogenetic mechanism responsible for the erythroid aplasia that develops in the course of a number of viral diseases, such as infectious mononucleosis and viral hepatitis (Wilson et al., 1980; Socinski et al., 1984). In a case of PRCA that developed in the course of transfusion-related hepatitis the patient's marrow was found to have a normal number of CFU-E. Whether there was an increased number of marrow lymphocytes or not was not reported. Addition to the culture media of the patient's peripheral blood lymphocytes suppressed normal and autologous marrow CFU-E development into erythroblastic colonies in vitro. The inhibition of erythroid colony formation was much more pronounced for normal (71 percent inhibition) than for autologous CFU-E (39 percent inhibition). The effect of these lymphocytes on autologous and normal CFU-GM was not reported, nor was the effect of lymphocytes from patients with hepatitis without PRCA on autologous CFU-E growth in vitro. Whether the lymphocyte-mediated suppression of normal erythropoiesis in vitro, present during the active phase of PRCA but undetectable after remission, played any major role in the pathogenesis of this syndrome in this case is difficult to assess in the absence of appropriate controls and documentation of significant marrow lymphocytosis. Whether it represents an epiphenomenon or not is not clear as the authors have also discussed (Wilson et al., 1980). A case of erythroid hypoplasia that developed in the course of chronic Epstein-Barr (EB) virus-induced infectious mononucleosis has also been studied by in vitro culture methods (Socinski et al., 1984). In this case the patient's marrow contained about 17 percent lymphocytes with a normal OKT 4:OKT 8 ratio; when whole marrow mononuclear cells were assayed in vitro for CFU-E an almost normal number of these progenitors was detected. After removal of T cells from the marrow the number of CFU-E grown in vitro increased almost threefold. Addition to T cell–depleted marrow cells of bone marrow but not peripheral blood lymphocytes resulted in a progressive decline of the CFU-E growth in vitro. Maximum inhibition was observed at a ratio of 1:1 of T-marrow lymphocytes to marrow cells. At a ratio of lymphocytes to marrow cells of 1:5, a ratio corresponding to the actual percentage of lymphocytes in the patient's marrow, the inhibition, although reported statistically significant, was most likely within the range of variation of the CFU-E assay. These findings, nevertheless, suggested that a bone marrow lymphocyte cell population in this case was probably exerting a suppressive role on the CFU-E development that might have been responsible for the erythroid hypoplasia. Whether these findings were specific for erythroid colony formation is not known, since the effect of marrow lymphocytes on CFU-GM was not studied. In addition, because infectious mononucleosis results in the generation of an increased number of activated suppressor T cells (Tosato et al., 1979), and because activated lymphocytes can suppress erythropoiesis and granulocytopoiesis in vitro (Bacigalupo et al., 1981; Banisadre et al., 1981; Harada et al., 1986), the association of these findings to the pathogenesis of erythroid hypoplasia is not straightforward, since similar studies on patients with infectious mononucleosis but without red cell aplasia have not been reported.

The design and interpretation of coculture experiments of lymphocytes with marrow cells to demonstrate a T cell–mediated suppression of erythropoiesis in cases of pure red cell aplasia is difficult. It is not easy to separate transfusion-induced lymphocyte effects from true autoimmune phenomena nor to demonstrate the specificity of the findings and their close relation to the pathogenesis of the erythroid aplasia. Torok-Storb and co-workers have addressed a similar problem in the study of T cell–mediated suppression of hematopoiesis in aplastic anemia and have proposed certain criteria for differentiating the transfusion-induced sensitization phenomena from findings suggestive of a possible autoimmune pathogenesis (Torok-Storb et al., 1980). In the case of isolated erythroid aplasia we feel that the following criteria should be fulfilled before any conclusion is made regarding a cell-mediated suppression of erythropoiesis of possible pathogenetic significance:

1. Growth of autologous erythroid colony-forming units from unseparated bone marrow cells is abnormally low as compared to the mean growth from a substantial number of normal controls.
2. Growth of erythroid colonies in vitro increases after removal from marrow cells of lymphocytes, and the increase is far beyond any increase that can be accounted for by enriching the marrow cell suspension in erythroid progenitors and significantly higher than the increase in the growth of granulocytic-monocytic progenitors.
3. Addition of lymphocytes to lymphocyte-depleted marrow cells, at a ratio equal to that present in unfractionated marrow cells, results in significant (probably greater than 40 percent) supression of erythroid but not myeloid colony formation in vitro.
4. The inhibitory activity of lymphocytes on erythropoiesis correlates with the activity of the disease.
5. In cases of secondary PRCA, lymphocytes from marrow of patients with the primary disease but without erythroid aplasia do not exert similar effects on autologous erythroid colony formation in vitro.
6. In cases where the lymphocytes originate from the blood and not the patient's marrow, it is shown that the marrow and blood lymphocytes have identical surface markers and the presence of increased lymphocytes with similar markers is documented on bone marrow biopsy or marrow particle sections.

Although a significant amount of evidence exists to suggest a primary role for T-lymphocytes in suppressing erythropoiesis in erythroid aplasia associated with chronic lymphocytic leukemia of B or T cell type and possibly in certain viral illnesses, the role of T-lymphocytes in the pathogenesis of primary acquired pure red cell aplasia remains unknown. In a preliminary study of five cases of chronic acquired primary PRCA, depletion of patients' marrow cells of T-lymphocytes resulted in no increase or change of the erythroid colony formation in vitro (Dessypris and Krantz, unpublished observations). Further studies are necessary to define the role of T cells in the pathogenesis of primary acquired PRCA and the mechanisms of T cell–mediated suppression in cases associated with lymphoproliferative and other disorders.

Human Parvovirus-like Infections and Aplastic Crises of Chronic Hemolytic Anemias

The infectious etiology of the aplastic crises seen in chronic hemolytic anemias has been suspected for a long time, not only because patients frequently provide a history consistent with a preceding viral illness, but also because of the clustering of cases in time and within families affected by a congenital type of hemolytic anemia. The association, however, of a human parvovirus-like infection (SPLV) with the acute transient erythroblastopenia seen in chronic hemolytic anemias and with the fifth disease (erythema infectiosum) was established only recently (Anderson 1982; Young and Mortimer, 1984).

The presence of SPLV in human sera was first described by Cossart and co-workers in 1975. These investigators were screening donated units of blood for the hepatitis B surface antigen (HBsAg) by counter-immunoelectrophoresis using a human serum as a source of antibody. They noticed that a number of blood samples that gave positive results by the above method for detection of HBsAg were found to be negative when tested by the more sensitive and specific passive hemagglutination method. Examination of the serum from these blood units by electron microscopy demonstrated the presence of uniform spherical particles 20–23 nm in diameter different from the pleiomorphic particles of hepatitis B. Although initially it was thought that this virus may be related to non-A, non-B hepatitis, studies in the general population and in patients with this type of hepatitis showed that the frequency of detection of antibodies in both groups is very similar (about 30 percent) (Cossart et al., 1975; Paver and Clark, 1976), making it an unlikely candidate for the non-A, non-B hepatitis. Until 1981, the SPLV was not associated with any specific clinical syndrome, other than a nonspecific viral illness (Cossart et al., 1975: Schneerson et al., 1980; Anderson, 1982), and its final association with the aplastic crisis of chronic hemolytic anemias was due to plain chance. Out of 800 serum samples taken from children with acute illness and examined by the virology laboratory at the King's College Hospital in London only two were found to react with serum containing antibodies to SPLV particles. Further investigation of these two cases revealed that both sera came from children with sickle cell disease who presented in aplastic crisis (Anderson, 1982). Active infection with this virus was subsequently reported in five children with sickle cell anemia and acute transient erythroblastopenia (Pattison et al., 1981). In a study of 112 children with sickle cell anemia who developed an aplastic crisis during the period 1952–80, it was noted that outbreaks occurred in 1956, 1960, 1965–67, 1971–73, and 1979–80. Cases of aplastic crisis in children over the age of 15 were rare. Sera from 28 out of 38 cases in the 1979–80 outbreak were stored and examined for the presence of either SPLV particles or antibodies to SPLV. In 24 out of 28 cases there was evidence for infection with SPLV, whereas only 8 out of 48 control children with normal hemoglobin were found to have antibody to SPLV (Serjeant et al., 1981). Further studies have subsequently documented the association of SPLV infection with the aplastic crisis seen in children not only with sickle cell anemia (Pattison et al., 1981; Serjeant et al., 1981; Anderson et al., 1982; Rao et al., 1983), but also in hereditary spherocytosis (Kelleher et al., 1983), thalassemia (Rao et al., 1983), and pyruvate kinase deficiency (Duncan et al., 1983), whereas sera from normal children with transient erythroblastopenia of childhood or from adult patients with acquired PRCA, aplastic anemia, or paroxysmal nocturnal hemoglobinuria were negative for evidence of recent SPLV infection (Young et al., 1984*b*). Following the epidemiologic association of SPLV with the aplastic crisis of chronic hemolytic anemias, the same virus was found to be the cause of erythema infectiosum (fifth disease) during an epidemic of this exanthematous disease of childhood among children in London (Anderson et al., 1982).

Human serum parvovirus is a DNA virus that probably will be classified in the genus of parvoviridiae, a group of viruses mainly known for causing a variety of diseases in vertebrate animals (Tattersall and Ward, 1978). These viruses consist of single-stranded DNA and are capable of replicating only in highly proliferating cells. The SPLV has characteristic morphology and size and is easily recognizable by electron microscopy. The mode of transmission of this virus in not yet fully established but it seems that the alimentary or respiratory tract as well as the hematogenous route may serve as portals for its entry (Anderson, 1982; Mortimer et al., 1983*b*). Evidence exists that the virus detected in human serum is different from the one detected in feces (Paver and Clarke, 1976). After infection with SPLV a short period of viremia follows; then a classic immune response, characterized initially by IgM and later by IgG specific antibodies, takes place. It seems that immunity is acquired during childhood and by the age of 15 about 30 to 50 percent of children have detectable antibodies. This may provide an explanation of the fact that aplastic crisis is rarely, if ever, seen for a second time in the same individual. Infection of a susceptible patient with chronic hemolytic anemia does not necessarily lead to acute erythroblastopenia. As noted by Anderson and colleagues (1982) as well as other investigators (Serjeant et al., 1981), not all patients with sickle cell disease with antibodies to SPLV detectable in their sera had a history of aplastic crisis, and seven patients with sickle cell disease seroconverted during the two-year study period without developing aplastic crisis. It seems that it is not only the infection by the SPLV but the proliferative status of the erythroid cells in the patient's marrow as well that determines the clinical appearance of aplastic crisis. The common finding in those patients who seroconverted without developing an aplastic crisis and without altering their reticulocyte count was a history of recent red cell transfusions. It is conceivable that transfusion of red cells decreased the level of circulating erythropoietin and consequently the CFU-E pool size in the marrow and their proliferative status was lowered significantly prior to exposure to the virus.

The pathogenesis of virus-induced aplastic crisis has been investigated in a number of studies (Duncan et al., 1983; Mortimer et al., 1983*a*; Young et al., 1984*a*, 1984*b*). Serum from patients with aplastic crisis contains viral particles. This serum when added to normal bone marrow cell cultures inhibits the growth and development of the erythroid progenitors into mature erythroblasts. This effect of virus-containing sera can be neutralized by mixing it with sera from the convalescent phase of the disease, which presumably contain antibodies to SPLV. It seems that the virus affects the growth and development of the late erythroid progenitor CFU-E much more than it does the early erythroid progenitor BFU-E. This selective effect of the virus on the growth of CFU-E rather than BFU-E was much more pronounced when a highly purified preparation of the virus was used in these studies. It should be noted, however, that this finding was obtained by studying the effect of virus on normal marrow cells in which there is a substantial difference between the percentage of CFU-E and BFU-E in DNA synthesis (70–90 percent versus 20–30 percent, respectively). It is possible that in the marrow of patients with chronic severe hemolytic anemia due to high levels of erythropoietin the BFU-E might be in a much more active proliferative state than in the normal marrow since erythropoietin can induce them into DNA synthesis (Dessypris and Krantz, 1984). Alternatively the virus may affect only the CFU-E because of the presence on its surface of certain differentiation-related antigens (receptors) that might serve as sites for its attachment. In another study, however, virus-containing serum reduced by 79 percent the numbers of blood BFU-E compared to that found in normal or convalescent sera (Duncan et al., 1983). In all studies the virus had little or no effect on the growth of CFU-GM. There seems to be a preferential involvement of erythroid rather than myeloid progenitors, although the latter were found to be mildly reduced in experimentally induced infection (Potter et al., 1987). The mild decrease in the total white cell and the absolute neutrophil counts observed in both natural and experimentally induced SPLV infection (Anderson et al., 1985; Potter et al., 1987) cannot be attributed to depression of marrow myeloid progenitors and precursors. Severe granulocytopenia has been thus far described only in a single case of persistent SPLV infection in an immunocompromised host (Kurtzman et al., 1987). The mechanism through which SPLV causes loss of proliferative capacity of CFU-E has been demonstrated by the studies of Young et al. (1984*a*). It seems that CFU-E infected by the virus are undergoing morphologic changes consistent with a direct toxic effect of the virus on these cells. Using a monoclonal antibody to SPLV, specific fluorescence was shown only in a minority of cells 24 to 48 hours following infection in vitro, and parvovirus-like particles were detected in arrays in the nucleus by microscopy. The SPLV (B19 parvovirus) has been successfully propagated in suspension cultures of human erythroid marrow cells from patients with hemolytic anemia. In this culture system newly synthesized viral particles were released from infected erythroid cells into the supernatant medium. The replication of the parvovirus in vitro was dependent on the presence of a high proportion of erythroid cells among the marrow cell suspension and on the addition of erythropoietin (Ozawa et al., 1986). The toxic effect of virus on CFU-E is independent of any marrow accessory cell or the presence of immunoglobulins and complement. These findings suggest that SPLV affects this erythroid progenitor directly and exerts a toxic effect upon it. In normal individuals cessation of erythropoiesis for a few days to a week is not expected to result in severe anemia but there may be a 5–10 percent temporary drop of the red cell mass that may pass unnoticed. However, in patients with a reduced red cell survival and a high demand on the marrow to produce red cells, such an arrest of erythropoiesis, no matter how short, may result in acute profound exacerbation of the chronic anemia.

Although the discovery of SPLV as a cause of aplastic crisis has provided answers to a number of questions, a number of issues regarding the interaction of the virus with the erythroid cells remain unanswered. The discovery of SPLV as a cause of aplastic crisis in children with hemolytic anemia has opened a new area of investigation, that of virus-induced bone marrow injury.

Erythroid Aplasia in Thymoma

The pathogenesis of selective erythroid aplasia in patients with thymoma has been addressed by numerous investigators. It was initially proposed that the thymic tumor secretes a humoral inhibitory factor that selectively suppresses erythropoiesis. Attempts to induce anemia and reticulocytopenia in laboratory animals by injecting them with saline extracts of the thymic tumor have been repeatedly unsuccessful (Clarkson and Prockop, 1958; Roland, 1964), and in a single case in which such an inhibition was demonstrated, the thymic extract contained a substantial amount of IgG that was found to have potent inhibitory activity on erythropoiesis in mice (Jepson and Vas, 1974). Many studies after the above initial reports have shown that the serum of patients with thymoma and PRCA contains a specific erythropoietic inhibitor localized in the IgG fraction (Barnes, 1965, 1966; Jepson and Lowenstein, 1966; Field et al., 1968; al-Mondhiry et al., 1971; Peschle and Condorelli, 1972; Zalusky et al., 1973; Marmont et al., 1975; Lowenberg and Ghio, 1977). The action of this inhibitor on in vivo and in vitro erythropoiesis has been already discussed (see chapter 3). Again, it should be noted that all these studies were performed on either murine or heterologous human marrow cells and they should be interpreted with caution considering all the problems inherent in such systems. The frequency of such inhibitors in the sera of patients with thymoma and PRCA is not known.

Suppression of normal erythropoiesis by peripheral blood lymphocytes has been described in two patients with thymoma; however, there was no correlation of in vitro findings and the presence or absence of erythroid aplasia (Litwin and Zanjani, 1977). In a more recently studied case of PRCA associated with thymoma and panhypogammaglobulinemia, the patient's marrow was found to be infiltrated by lymphocytes that constituted 33 percent of the nucleated marrow cells. Almost all of these lymphocytes expressed surface markers consistent with a population of activated suppressor-cytotoxic T cells (OKT11+, OKT3+, OKT8+, Ia+). Erythroid colony and burst formation from the patient's whole marrow mononuclear cells was very low, but increased almost tenfold after depletion of marrow cells of T-lymphocytes by the sheep erythrocyte rosetting technique. Addition of marrow T cells to T cell–depleted marrow cells at a ratio of 1:4, a ratio not exceeding the percentage of T cells among marrow cells, resulted in significant suppression of erythroid colony formation in vitro and the suppression was found to be dependent on the dose of added T-lymphocytes. Following treatment of the patient with prednisone and cyclophosphamide and recovery of erythropoiesis, the percentage of T-lymphocytes among marrow cells decreased to 8 percent and these T cells no longer had any effect on erythroid colony formation by the patient's marrow cells in vitro.

In this study depletion of marrow cells of T cells by treatment with OKT3 monoclonal antibody and complement had a surprisingly small effect on erythroid colony formation and the effect of T cell depletion on the growth of granulocyte-macrophage progenitors in vitro was not reported

(Mangan et al., 1986). As in the case studied by Lipton et al. (1983) T-lymphocytes in this case also had a suppressive effect on autologous, but not normal, marrow erythroid colony formation in vitro. The findings in this study demonstrated that in this patient erythroid aplasia was most likely mediated through the action of T-lymphocytes on marrow erythroid cell development. The role of T-lymphocytes in mediating the erythroid aplasia in patients with thymoma needs further investigation.

The mechanism through which a thymoma induces a variety of autoimmune disorders (Souadjian et al., 1974) is poorly understood. Studies of the immune function in patients with thymoma are lacking. The coexistence of immunodeficiency and autoimmune diseases in patients with thymoma indicates that a profound abnormality of regulatory immune mechanism(s) is present that eventually leads to failure of thymus-dependent immune surveillance. More recent studies have suggested that abnormalities of the secretory function of the endocrine thymus (Trainin, 1974; Schulof and Goldstein, 1977) are associated with a number of immunodeficiency, autoimmune, and neoplastic diseases (Bach and Dardenne, 1972; Wara et al., 1975; Schulof and Goldstein, 1977). Elevated levels of thymosin-a_1 have been recently reported in two patients, one with a malignant thymoma (Chollet et al., 1981) and another with possible thymoma (detectable only microscopically), PRCA, and myasthenia gravis (Socinski et al., 1983). This polypeptide hormone has been localized in thymic epithelial cells, which are considered to be the origin of thymomas (Dalakas et al., 1981; Lauriola et al., 1981). The possible contribution of thymic hormones produced by neoplasms of thymic epithelium in the pathogenesis of the autoimmune and immunodeficiency disorders so commonly present among patients with thymomas is not yet known.

Autoimmune Hemolytic Anemia and Pure Red Cell Aplasia

Reticulocytopenia in the course of immune hemolytic anemia may be associated either with erythroid aplasia or with erythroid hyperplasia. Since both autoimmune hemolytic anemia and PRCA are associated with the presence of autoantibodies, it was suggested that antibodies directed against erythrocytes may in certain cases also attack the erythroid precursors in the marrow and cause erythroid aplasia (Gasser, 1949; Rohr, 1949; Eisemann and Dameshek, 1954). It was initially demonstrated by the use of the antiglobulin reaction that the nucleated erythroid precursors in a number of patients with autoimmune hemolytic anemia were coated with red cell autoantibodies (Steffen, 1955; Pisciotta and Hinz, 1956). However, coating of the marrow erythroid precursors with autoantibodies does not seem to be sufficient for the production of erythroid aplasia. It was subsequently shown that erythroblasts in the marrow of patients with autoimmune hemolytic anemia, when incubated with the patient's serum, may develop morphologic signs of cell injury (Rossi et al., 1957). Since marrow erythroblasts express on their membranes a number of antigens present on the red cells (Yunis and Yunis, 1963, 1964) it is not surprising that autoantibodies can be detected on marrow erythroblasts, or that erythroid aplasia occurs only in those cases where the autoantibody reacts with the erythroid progenitors themselves. The latter has been supported by more recent work by a number of investigators. Meyer and co-workers (1978) studied a case of autoimmune hemolytic anemia and periodic pure red cell aplasia in a patient with systemic lupus erythematosus and demonstrated that the patient's serum as well as the IgG eluted from his red cells, but not from the red cells of other patients with autoimmune hemolysis without PRCA, were capable of suppressing the growth of normal marrow CFU-E in vitro. The results of these experiments indicated that in the rare patient with these two autoimmune disorders the reactivity of the autoantibody may include the whole spectrum of erythroid cells from the CFU-E all the way down the differentiation pathway to the red cell. We also had the opportunity to study a patient with mixed connective tissue disorder who developed autoimmune hemolytic anemia, PRCA, and immune thrombocytopenic purpura. An IgG panagglutinin was detected on the patient's erythrocytes and in her plasma by the antiglobulin test. This patient's marrow CFU-E and BFU-E were normal in numbers as assayed by the plasma clot system. Addition to the culture medium, which was rendered complement-free by heat inactivation, of the patient's purified IgG or IgG eluted from her red blood cells had no effect on the erythroid colony formation in vitro. However, preincubation of the patient's marrow cells with serum IgG or red cell eluate with fresh AB serum as a source of complement, followed by washing of the marrow cells, resulted in a 70–90 percent decline in the number of autologous marrow CFU-E growth in vitro. The substitution of fresh for heat-treated serum abolished the inhibitory effect of both the serum IgG and the red cell eluate (table 9). No effect could be detected on CFU-GM. It is notable that the eluate was 30 times more potent as an inhibitor than the serum IgG on a mg per mg of protein basis. These findings are in accordance with the results of the above-mentioned study by Meyer and colleagues, and demonstrate in an autologous system that in certain cases of PRCA and autoimmune hemolytic anemia the autoantibody may recognize antigenic determinants common to the CFU-E and the mature erythrocyte (Dessypris et al., unpublished observations).

In another case studied by Mangan et al. (1984), the antierythrocyte autoantibody had an anti-ee specificity and was complement-independent, whereas the patient's serum contained an IgG that in a complement-dependent fashion was capable of inhibiting autologous and normal CFU-E and late BFU-E growth in vitro. The eluate from ee-negative red cells had

Table 9 The Effect of Serum IgG and Red Cell Eluate on the Growth of Autologous Marrow CFU-E in a Patient with Pure Red Cell Aplasia and Autoimmune Hemolytic Anemia

	CFU-E/1 × 10⁵ plated cells*
CFU-E assay in a complement-free system⁺	
Medium alone	418 ± 53
Normal IgG (1mg/ml)	498 ± 44
Patient's IgG (1mg/ml)	444 ± 61
RBC-eluate (0.1mg/ml)	513 ± 63
Preincubation of marrow cells for 1 hour at 37° C**	
Medium	198 ± 22
Medium + 50% fresh AB serum	231 ± 19
Normal IgG	256 ± 31
Normal IgG + 50% fresh AB serum	203 ± 27
Patient's IgG	193 ± 18
Patient's IgG + 50% fresh AB serum	62 ± 11
RBC eluate	247 ± 26
RBC eluate + 50% fresh AB serum	31 ± 4
RBC eluate + 50% heat-inactivated serum	279 ± 31

⁺ Assay performed in plasma clots
* Mean ± sem of quadruplicates
** After preincubation cells were washed twice with medium resuspended in their initial volume of medium and assayed in plasma clots.

no effect on the erythroid colony formation in vitro. These studies demonstrated that in this case of PRCA and autoimmune hemolytic anemia two different autoantibodies were responsible for the clinical picture, one acting on red cells and the other on the patient's marrow erythroid progenitors. It seems, therefore, that in the PRCA associated with immune hemolysis these may be either one responsible autoantibody attacking both red cells and erythroid progenitors, or two different autoantibodies with distinct specificities.

Autoimmune hemolytic anemia with reticulocytopenia and erythroid hyperplasia in the marrow has been repeatedly reported (Crosby and Rappaport, 1956; Harley and Dods, 1959; Buchanan et al, 1976; Celada et al., 1977; Hedge et al., 1977; Seewann, 1979; Conley et al., 1980, 1982; Hauke et al., 1983). The mechanism responsible for the reticulocytopenia was proposed to be a preferential destruction of reticulocytes (Hegde et al., 1977). However, this hypothesis was not confirmed by a more recent study (Conley et al., 1982). In at least two cases of this syndrome, ferrokinetic studies of the erythron have demonstrated the presence of ineffective erythropoiesis (Seewann, 1979; Conley et al., 1982). Since ineffective erythropoiesis with hyperplastic erythroid marrow has been recognized as part of the natural history of erythroid aplasia (see Bone Marrow Morphology in chapter 2), it is reasonable to suggest that these cases of reticulocytopenia and erythroid marrow hyperplasia in the course of autoimmune hemolytic anemia may indeed represent a phase of PRCA associated with autoimmune hemolysis. Hauke and colleagues (1983) studied a case of reticulocytopenia with erythroid hyperplasia associated with immune hemolytic anemia and demonstrated that the patient's plasma inhibited in a dose-dependent fashion the growth of normal marrow BFU-E in vitro, an effect absent from the same patient's plasma after a splenectomy-induced remission. This effect was not demonstrable on normal marrow CFU-GM or CFU-GEMM (multilineage colony-forming units with granulocytes, erythroblasts, monocytes, and megakaryocytes). Although the significance of these findings is not well understood due to the fact that the experiments were performed not in a completely autologous system (patient's plasma and bone marrow cells); nevertheless, they suggest that the above-raised hypothesis may indeed be true. Further studies in completely autologous systems are needed in similar cases to further investigate this hypothesis.

The mechanism through which an autoantibody to erythroid progenitors and red cells may result in erythoid hyperplasia and reticulocytopenia is not known. It is conceivable, however, that injury to the erythroid progenitors may not necessarily lead to death of these cells and erythroid aplasia, but it may be expressed on more mature cells derived from these progenitors in the form of ineffective erythropoiesis. Furthermore, immune suppression of erythropoiesis either may lead to full-blown PRCA or may exist in a low-grade compensated form (Dessypris et al., 1984), and clinical syndromes in between these two extreme points may be recognized in the future.

Collagen Vascular Diseases and Pure Red Cell Aplasia

Isolated erythroid aplasia has been reported in association with rheumatoid arthritis (adult and juvenile) and with systemic lupus erythematosus (see chapter 1). The pathogenesis of this hematologic complication in the course of rheumatoid arthritis and systemic lupus erythematosus has been addressed in a number of studies. In a case of PRCA in the course of SLE studied by Cavalcant et al. (1978), the number of mature (day 11) BFU-E in the patient's marrow was found to be increased compared to normal marrow despite an almost complete absence of erythroblasts from the patient's marrow indicating that the arrest of erythropoiesis was not affecting the development of mature BFU-E from its earlier erythroid progenitor. The development of mature BFU-E in vitro was complete proceeding to the stage of colonies consisting of orthochromatic erythroblasts.

Addition of autologous blood lymphocytes had no effect on the erythroid colony formation in vitro. Although 5 percent of the patient's serum when added to the culture medium had no effect, the addition of purified serum IgG depressed the number of erythroid colonies formed in vitro to 30 percent of control normal IgG. This phenomenon of the absence of demonstrable inhibitor at a low concentration of serum (less than 5–10 percent) but the presence of inhibitory activity in the purified IgG fraction (1–2 mg/ml) also has been seen in our laboratory in a number of cases. The studies in this case were performed in a completely autologous system and indicate that PRCA in SLE may be caused by an IgG inhibitor of erythropoiesis.

In a case of PRCA in a patient with longstanding rheumatoid arthritis, an assay of the patient's marrow during the period of erythroblastopenia showed an extremely poor or absent growth of marrow CFU-E and BFU-E in vitro. The number of CFU-E and BFU-E in the patient's marrow returned, however, to normal after an azathioprine-induced remission of PRCA. At that time the in vitro assay of the patient's serum IgG for possible inhibitory activity demonstrated that IgG purified from serum collected during the active phase of PRCA suppressed both autologous CFU-E and BFU-E growth significantly in a dose-dependent fashion, but had no effect on the CFU-GM growth in vitro. IgG purified from serum drawn after the remission of PRCA at low concentrations (1–2 mg/ml) had no effect on erythroid colony development in vitro, but at high concentrations exerted some degree of inhibition the magnitude of which was only a small fraction of that observed by the same concentration of IgG from the active phase of the disease (figure 15). It seems that in this patient humoral inhibition of erythropoiesis was probably responsible for the production of erythroid aplasia (Dessypris et al., 1984). Similar studies performed in 20 patients with rheumatoid arthritis and anemia of varying severity not due to PRCA but to chronic inflammation have shown no inhibitory activity of serum IgG on autologous marrow erythroid colony formation (Baer and Dessypris, unpublished observations), indicating that the presence of an erythropoietic inhibitor in this patient with rheumatoid arthritis and PRCA was related to the presence of the PRCA rather than the rheumatoid arthritis per se. In this case also the observation that during remission of PRCA this IgG inhibitor was still present, but at much lower concentrations, in the patient's serum suggested that low-grade humoral inhibition of erythropoiesis may contribute to the moderate anemia so commonly seen in this disease. This was previously suggested by Dainiak et al. (1980), who reported the presence of serum inhibitors of erythropoiesis detected on normal, but not autologous, marrow cells in 10 out of 17 patients with rheumatoid arthritis, systemic lupus erythematosus, or vasculitis secondary to hepatitis B. We were unable, however, to reproduce these findings using

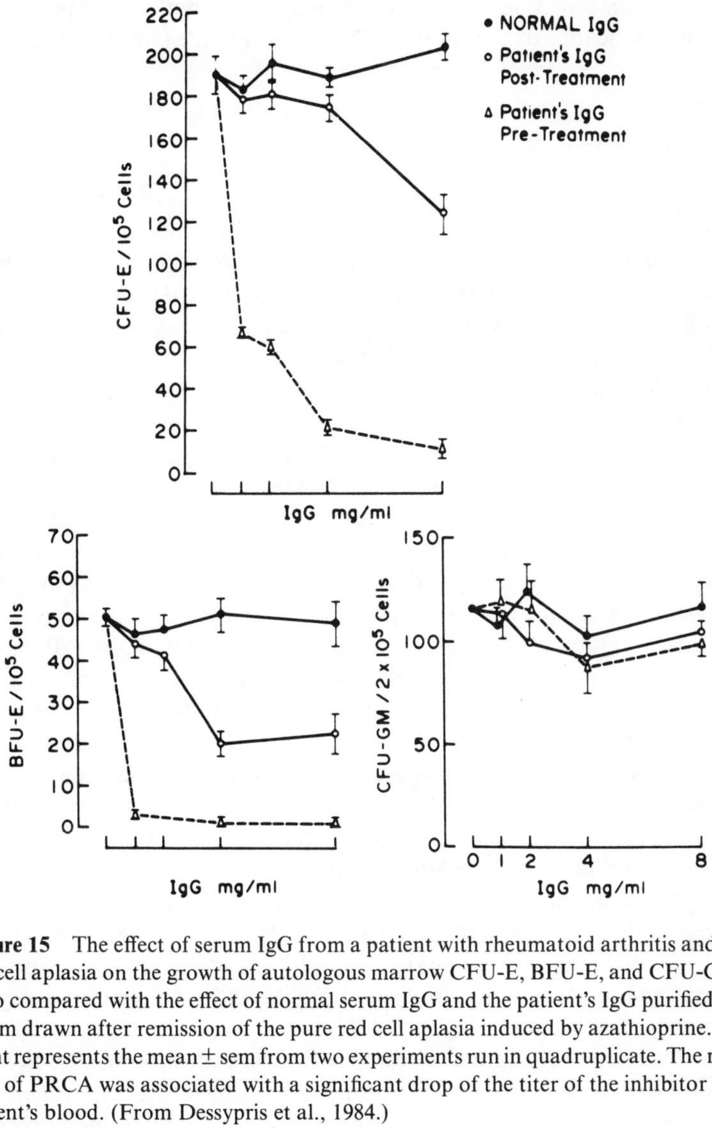

Figure 15 The effect of serum IgG from a patient with rheumatoid arthritis and pure red cell aplasia on the growth of autologous marrow CFU-E, BFU-E, and CFU-GM in vitro compared with the effect of normal serum IgG and the patient's IgG purified from serum drawn after remission of the pure red cell aplasia induced by azathioprine. Each point represents the mean ± sem from two experiments run in quadruplicate. The remission of PRCA was associated with a significant drop of the titer of the inhibitor in the patient's blood. (From Dessypris et al., 1984.)

autologous IgG and marrow or blood cells in a similar group of patients (Baer and Dessypris, unpublished observations).

In another case of PRCA and rheumatoid arthritis, T cell inhibition of erythropoiesis was suggested by the fact that depletion of the patient's marrow of T-lymphocytes resulted in a 90 pecent increase of the CFU-E growth in vitro. However, in this study normal marrow cells when depleted of

T-lymphocytes gave rise to 70 percent more CFU-E, making the pathophysiologic significance of the results highly questionable (Konwalinka et al., 1983).

Drug-induced Pure Red Cell Aplasia

The pathogenesis of drug-induced PRCA has thus far been addressed only in a limited number of cases of diphenylhydantoin-or isoniazid-induced erythroid aplasia. Basically two hypotheses have been raised regarding the drug-induced PRCA; a direct toxic effect of the responsible drug on erythroid cells resulting in selective aplasia of the erythroid line and an autoimmune mechanism.

The first case of PRCA that was reported in association with diphenylhydantoin treatment (Brittingham et al., 1964) was further investigated and reported by Yunis et al. (1967). In this case rechallenge of the patient with diphenylhydantoin resulted in induction of erythroid aplasia that remitted soon after the discontinuation of medications. This experiment was repeated five times with the same result. The fact that reinduction of erythroid aplasia could be achieved with small doses of this drug was interpreted as supportive of a possible immunologic mechanism. To further investigate this possibility a normal volunteer was given diphenylhydantoin and was infused with 500 ml of plasma, collected from the patient during the active phase of the aplasia, daily for three days. The recipient was followed with serial determinations of the reticulocyte count, number of marrow erythroblasts, serum iron, iron binding capacity, and ^{59}Fe clearance. These parameters remained unchanged during the follow-up period, which led these investigators to the conclusion that the erythroid aplasia was not immunologically mediated. In further studies in vitro they demonstrated that diphenylhydantoin when added to the patient's marrow cells suspended in his own plasma was capable of inhibiting DNA synthesis. They concluded that the toxic effect of diphenylhydantoin in this patient was exerted through inhibition of DNA synthesis by marrow erythroid precursors (Yunis et al., 1967).

Another case of diphenylhydantoin-induced PRCA was studied by Lee and associates (1978). These investigators, using the diffusion chamber technique for growing erythroid cells in vitro, were able to demonstrate that the patient's serum contained a toxic factor capable of reducing the number of erythroblast foci. The toxic factor disappeared from the patient's serum after remission of PRCA (Lee et al., 1978). An increased sensitivity of marrow erythroblasts to diphenylhydantoin and a direct toxic effect on erythroblasts as a mechanism for induction of erythroid aplasia was proposed by Sugimoto and co-workers (1982). These investigators showed that heme synthesis in vitro by the patient's marrow erythroblasts was inhibited in the presence of 25 μg/ml of diphenylhydantoin added to the culture medium. However, this concentration is toxic in vitro for all hematopoietic cells.

We had the opportunity to study the mechanism of diphenylhydantoin-induced PRCA in a 32-year-old man who developed this hematologic complication while on treatment with this drug, and in whom PRCA remitted within a few weeks following withdrawal of this medication. In our initial experiments performed on normal marrow cells, diphenylhydantoin was found to be toxic for normal marrow CFU-E, BFU-E, and CFU-GM in vitro, when added in concentrations exceeding 3μg/ml. Since therapeutic serum levels generally range from 10–20 μg/ml, the possibility that a number of individuals may develop PRCA because of an increased sensitivity to the toxic effect of the drug was considered unlikely. Furthermore, normal marrow myeloid progenitors were found to be at least twice as sensitive to the effect of diphenylhydantoin as normal erythroid progenitors. When the patient's IgG was purified from serum drawn during the active phase of PRCA and added to the culture medium it exerted no effect on the growth of normal marrow CFU-E or BFU-E in vitro. These IgG preparations were free of diphenylhydantoin by fluorescence polarization immunoassay and gas chromatography. Addition of subtoxic concentrations of diphenylhydantoin (1–2 μg/ml) to cultures containing the patient's serum IgG resulted in greater than 50 percent decline in the number of normal marrow CFU-E– and BFU-E–derived erythroid colonies but no change of the CFU-GM–derived granulocytic/monocytic colonies. When the same experiments were repeated with IgG purified from the patient's serum collected after remission of PRCA no effect could be demonstrated on erythroid colony formation in vitro (figure 16). These findings were further confirmed in a completely autologous system using the patient's blood as a source of erythroid progenitors. Autologous blood BFU-E growth in vitro was significantly suppressed by the patient's IgG only in the presence, but not in the absence, of diphenylhydantoin. In addition, this suppression of erythroid colony formation in vitro was demonstrable even when the patient's IgG and diphenylhydantoin were added to the culture medium after eight days of incubation, indicating that the target cell was an erythroid progenitor more mature than the blood BFU-E (figure 17). The complex of IgG and diphenylhydantoin was not inhibiting the response of normal marrow erythroblasts to erythropoietin in vitro, as assessed by measuring ^{59}Fe-incorporation into heme in vitro, it was not cytotoxic to ^{59}Fe-labeled erythroblasts, as assessed by the erythroblast cytotoxicity assay, and it was not cytotoxic to CFU-E, as assessed by assaying normal marrow mononuclear cells preincubated with this complex in the presence of complement. These studies suggested that in this patient erythroid aplasia was probably induced by an IgG antibody or immune

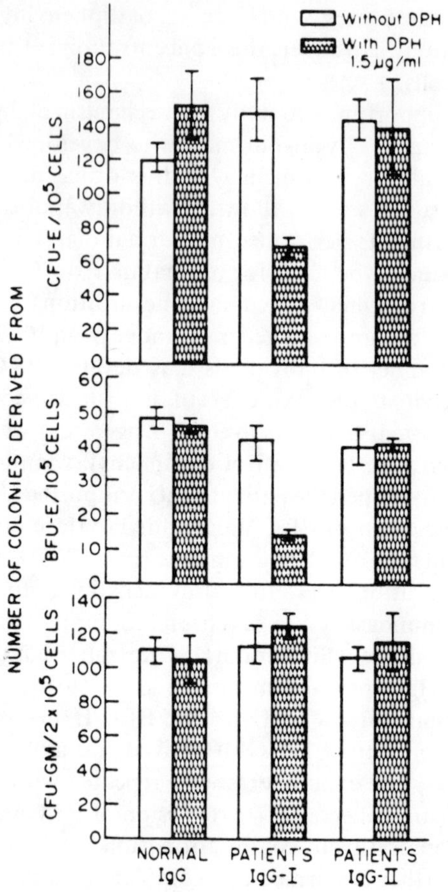

Figure 16 The effect of serum IgG from a patient with diphenylhydantoin-induced pure red cell aplasia on normal marrow CFU-E, BFU-E, and CFU-GM growth in vitro in the presence or absence of diphenylhydantoin (DPH). IgG-I denotes IgG purified from serum collected at the time of the diagnosis of PRCA and before any red cell transfusion was given. IgG-II denotes IgG prepared from serum collected after discontinuation of DPH and remission of PRCA. (From Dessypris et al., 1985.)

complex in cooperation with diphenylhydantoin, and that the target cell for this inhibitor was most likely an erythroid progenitor at the stage of differentiation between the CFU-E and the proerythroblast. This was further suggested by the fact that in cultures containing the patient's IgG and the drug an increased number of defective CFU-E colonies consisting of fewer than eight erythroblasts was seen as compared to control cultures. These findings could provide an explanation for the lack of effect of infu-

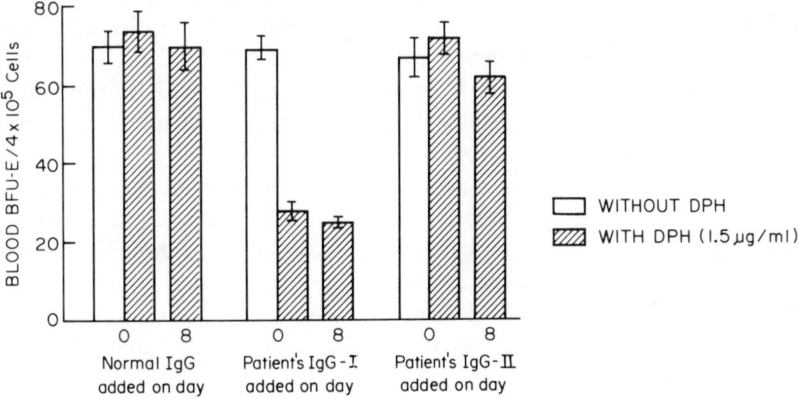

Figure 17 The effect of serum IgG from a patient with diphenylhydantoin-induced pure red cell aplasia on autologous blood BFU-E growth in vitro in the presence or absence of diphenylhydantoin (DPH). Suppression of autologous BFU-E growth was seen in the presence of IgG purified from serum collected at diagnosis (IgG-1) and DPH, but not in the presence of IgG purified from serum collected after remission of PRCA (IgG-2). The demonstration of suppression of autologous BFU-E growth upon the addition of IgG-I and DPH on the eighth day of culture suggests that the target cell for this complex is not the BFU-E per se but a BFU-E–derived, more differentiated cell. (From Dessypris et al., 1985.)

sion of the patient's serum on the erythropoiesis of the normal volunteer treated with diphenylhydantoin that was reported by Yunis and co-workers (1967). If the target cell for this inhibitory complex of IgG and diphenylhydantoin is an early cell before the stage of proerythroblast, one would have to infuse the volunteer with the patient's plasma to achieve and maintain adequate concentration of IgG for a period of at least five to seven days, needed to exhaust the marrow reserve of already differentiated erythroid precursors, in order to see any effect on the reticulocyte count, number of marrow erythroblasts, or ^{59}Fe clearance. This time period corresponds to the time needed for an erythroid progenitor to mature and develop into a reticulocyte.

The interaction between the patient's IgG and diphenylhydantoin is not known. Although it is conceivable that diphenylhydantoin formed an immune complex with the IgG, no diphenylhydantoin was detectable in the IgG preparation and we were unable to demonstrate any increased affinity of this drug for the IgG (Dessypris et al., 1985).

Hoffman and co-workers (1983) studied four patients with isoniazid-induced PRCA. In two of three patients the number of CFU-E assayed in the patients' marrow cells was found to be normal; the BFU-E in three out of three patients' marrows were normal. Addition to the medium for culturing autologous or normal marrow erythroid progenitors of the patient's serum with and without isoniazid at therapeutic concentrations had no effect on erythroid colony development in vitro. Whether the erythroid aplasia in these cases was immunologically mediated or whether it was the result of a direct toxic effect of the drug or its metabolites on the erythroid marrow cells is not known.

Cases of erythroid aplasia have been reported in association with treatment with chloramphenicol, thiamphenicol, and azathioprine. The direct toxicity of both chloramphenicol and its analogue, thiamphenicol, as well as of azathioprine on the erythroid marrow has been well described (McCurdy, 1961; Yunis and Bloomberg, 1964), and the relation of the magnitude of toxicity to the dose of the drug favors a direct toxic effect as the mechanism responsible for the induction of erythroid aplasia.

Transient Erythroblastopenia of Childhood

Tillmann et al. (1976) and Ritter and Zeller (1977) initially postulated that the anemia of TEC may be due to an arrest of erythropoiesis at the level of the erythroid-committed stem cell. The pathogenesis of isolated erythroid aplasia in previously hematologically normal children was first investigated by Wegelius and Weber (1978). They examined the sera of five children with TEC for the presence of antierythroblast antibodies. Their findings were negative. However, as they commented, this study was performed after recovery and in some cases several months after the actual damage to the red cell precursors. Koenig and colleagues (1979) studied the effect of serum IgG from four patients with TEC on erythroid colony formation in vitro by normal bone marrow cells. They reported that the mean number of CFU-E−, mature BFU-E−, and primitive BFU-E−derived colonies was statistically lower in the presence of the patient's IgG as compared to the number of erythroid colonies grown in the presence of normal IgG. In a single case they also tested the effect of the patient's serum on autologous marrow cells and showed that the patient's serum was highly inhibitory for the erythroid colony formation in vitro from autologous erythroid progenitors. They concluded that the anemia in TEC is due to transient immune suppression of erythroid colony formation. Katz and associates reported on the proliferative capacity of erythroid marrow cells in three cases of TEC. They found that the number of CFU-E in the marrow of these children was significantly reduced as compared to normal marrow and they postulated that the erythroid aplasia in these patients was probably the consequence of an erythroid stem cell defect or of pathologic cellular interactions with nonerythroid regulatory cells (Katz et al., 1981). The proliferative capacity of bone marrow erythroid progenitors has been addressed in three other studies. A total of sixteen cases of TEC have been studied thus far, and in ten out of sixteen the number of CFU-E was found to be normal, in the remaining six, reduced. In eleven out of sixteen cases, the number of BFU-E was found within the normal range and in the remaining five, reduced or absent (Inoue et al, 1980; Dessypris et al., 1982; Freedman and Saunders, 1983). It seems that the quantification of erythroid colony formation in vitro cannot by itself offer any clue as to the pathogenesis of TEC, since low growth can be interpreted either as evidence of an intrinsic stem cell defect or as evidence of in vivo injury to the erythroid progenitors that renders them incapable of proliferating and differentiating in vitro into mature erythroblasts. The latter explanation seems most likely in view of the transient nature of the disease and the absence of relapses following recovery.

In a group of twelve cases of TEC, we were able to demonstrate the presence of a serum inhibitor of erythroid colony formation in vitro in three out of five evaluable patients. In four other patients the growth of CFU-E and BFU-E in vitro was so poor that no conclusion could be drawn regarding the presence of an inhibitor in their sera. In two additional cases no inhibitory effects of the patient's serum on autologous marrow cells could be detected (figure 18). The inhibitory activity of the serum was shown to be localized in the IgG fraction and was active both on autologous and on normal marrow erythroid progenitors (figures 19 and 20). The inhibitory effect of the patient's serum IgG was limited only to erythroid cells, since no inhibitory effect was noted on normal marrow CFU-GM growth in vitro. In addition it seems that the presence of a serum IgG inhibitor of erythropoiesis correlated with the activity of the disease since the inhibitory activity of serum IgG progressively declined within two weeks of diagnosis and eventually disappeared after recovery of erythropoiesis. IgG purified from the sera of children with TEC collected from blood drawn one to three months after recovery exhibited no inhibitory activity against normal marrow erythroid colony formation in vitro (figure 21).

In an attempt to investigate the mode of action of this erythropoietic inhibitor, the patient's serum IgG that was found to be inhibitory for autologous and normal erythroid colony formation in vitro was tested in liquid suspension culture for its effect on heme synthesis by marrow erythroblasts in the presence of erythropoietin. Normal marrow erythroblasts responded to erythropoietin by synthesizing the same amount of heme in the presence of the patient's or normal IgG. In addition, in the erythroblast cytotoxicity assay, all tested sera IgG showed no toxicity against normal marrow erythroblasts. These findings suggested that in TEC the IgG inhibitor of erythropoiesis is not directed against the mature recognizable marrow erythroblasts and does not interfere with the action of erythropoietin on these erythroid cells. Since the major positive finding in these studies was the inhibition of erythroid colony formation in vitro, we further investigated

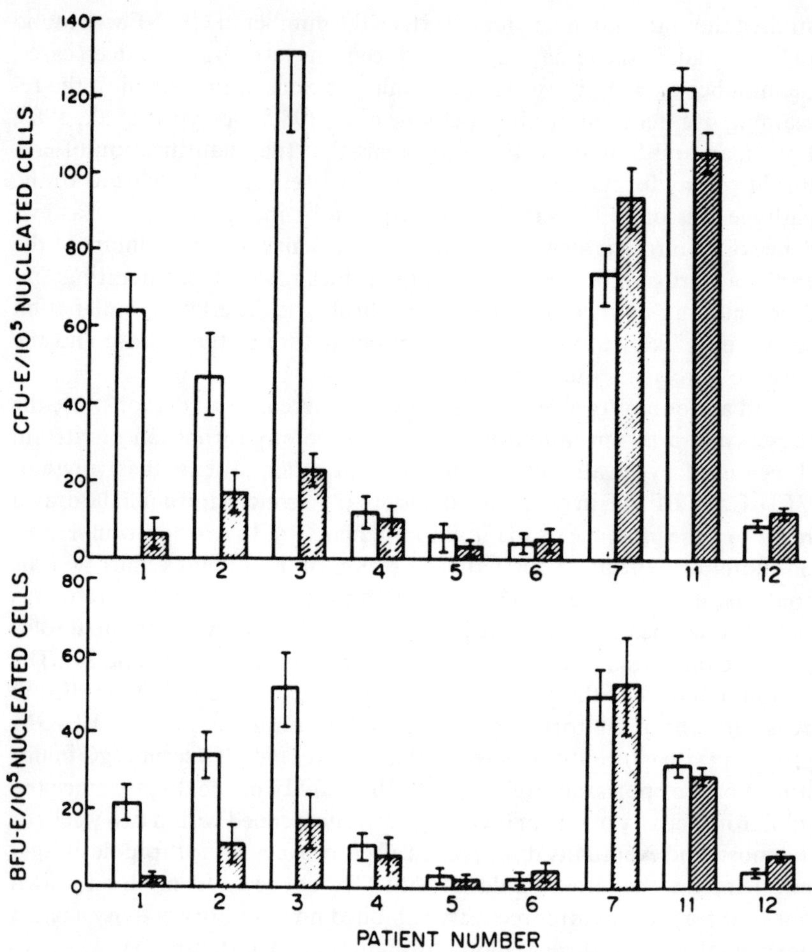

Figure 18 The effect of sera from patients with transient erythroblastopenia of childhood (shaded bars) compared with normal pooled serum (open bars) on the growth of autologous CFU-E and BFU-E. Mean ± sem of quadruplicates. TEC serum was added at a concentration of 10 percent (vol/vol). Suppression of CFU-E and BFU-E growth was observed in three out of five evaluable cases. (From Dessypris et al., 1982.)

the mechanisms by which such an inhibition takes place. By preincubating marrow cells containing erythroid progenitors with the patient's IgG with or without complement for one hour at 37° C, washing the IgG and complement away, and then assaying for CFU-E and BFU-E we were able to demonstrate that in four cases the IgG was directed against the CFU-E per se as opposed to erythroid cells that derive from the CFU-E. In three of these cases the IgG-induced injury to the CFU-E was complement-

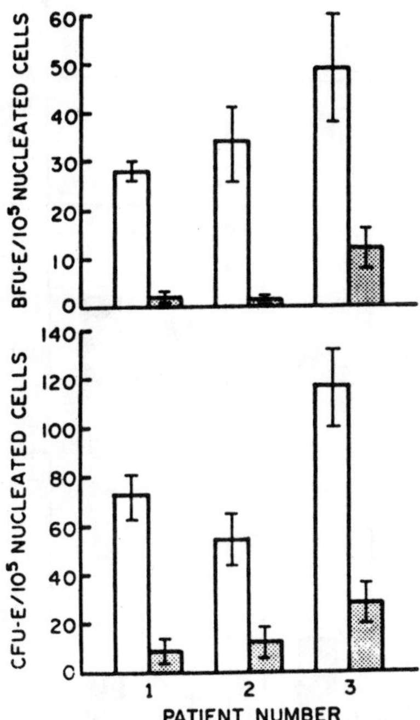

Figure 19 The effect of serum IgG from patients with transient erythroblastopenia of childhood on the growth of autologous marrow CFU-E and BFU-E in vitro. Mean ± sem of quadruplicates. Shaded bars represent cultures with patient's IgG. Open bars represent cultures containing normal pooled serum IgG at the same concentration. (From Dessypris et al., 1982.)

mediated and in the remaining one case complement-independent (figure 22). In the other two cases the inclusion of IgG in the culture medium suppressed CFU-E and BFU-E growth in vitro but a specific direct effect on the CFU-E or BFU-E per se could not be demonstrated. The results of all these studies are summarized in table 10 and indicate that in the majority of cases of TEC an IgG inhibitor of erythropoiesis in vitro can be identified. Its target cell seems to be either the early or late erythroid progenitor BFU-E or CFU-E, but not the recognizable marrow erythroid precursors. Its mode of action seems to be variable; in certain cases it acts as an antibody injuring the erythroid progenitors through a complement-mediated or independent mechanism, but in other cases it suppresses the development of erythroblasts from erythroid progenitors in vitro through an unknown mechanism (Dessypris et al., 1982).

In a group of four children with TEC, Freedman and Saunders (1983) were also able to demonstrate the presence of an IgG inhibitor of autolo-

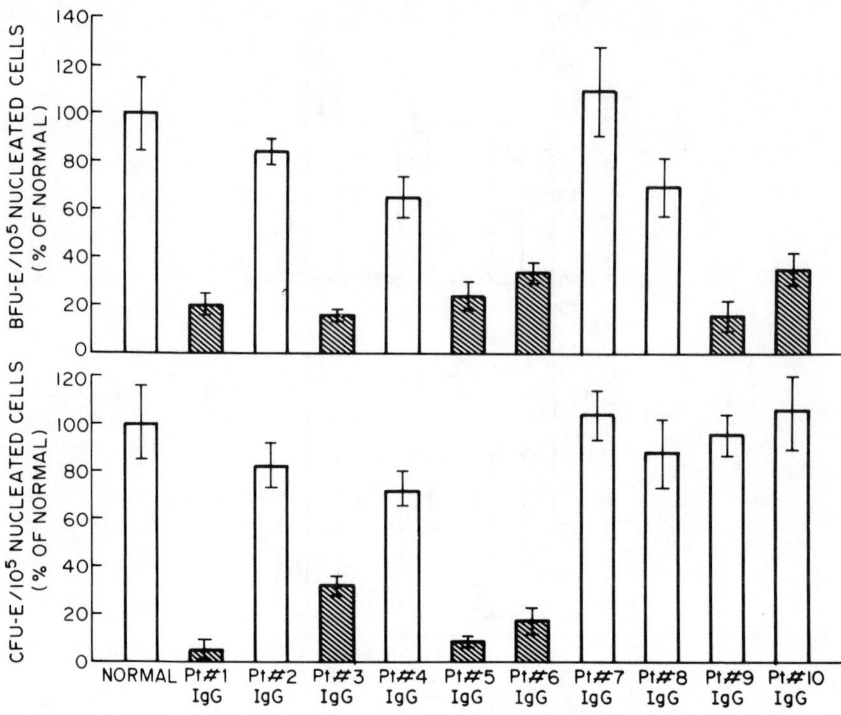

Figure 20 The effect of serum IgG from patients with transient erythroblastopenia of childhood on the growth of normal marrow CFU-E and BFU-E in vitro. Each bar represents the percentage of the number observed with normal IgG and corresponds to 183 $\pm$ 14 CFU-E/ 10^5 plated cells, and to 31 $\pm$ 4 BFU-E/ 10^5 plated cells. Shaded bars indicate cases with significant inhibition of growth. (From Dessypris et al., 1982.)

gous BFU-E growth in one patient, and an IgM inhibitor of autologous CFU-E and BFU-E growth in another case. In the third patient they attempted to investigate the role of the patient's peripheral blood mononuclear cells on the growth of normal marrow erythroid progenitors in vitro. They found a sharp decline of normal BFU-E but not CFU-E growth upon addition of the patient's mononuclear cells and concluded that in this case suppression of erythropoiesis was induced by the patient's mononuclear cells. Since the last study was performed on normal marrow cells it is not known whether this inhibition was the result of generation of toxic lymphocytes in a mixed lymphocyte culture (patient's and normal cells) or was due to the presence of a specific erythroid suppressor cell population. The role of peripheral blood mononuclear cells and purified blood T-lymphocytes on the erythroid colony formation by autologous bone mar-

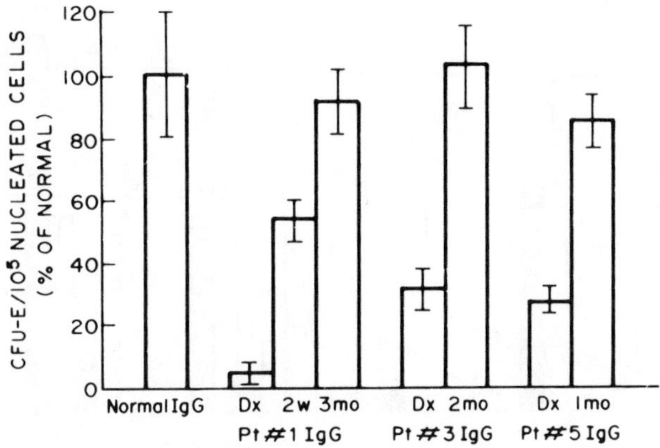

Figure 21 The effect of serum IgG, drawn at different stages of disease, from patients with transient erythroblastopenia of childhood on the growth of normal marrow CFU-E. The results are expressed as a percentage of the number observed with normal IgG ($134 \pm 17/10^5$ cells) and correspond to the mean ± sem of three different experiments. The inhibitory effect of the patients' IgG was parallel to the disease activity and was no longer present in the sera of patients in remission. Dx = serum drawn at the time of diagnosis before any red cell transfusion was given; 2w = serum drawn two weeks later, while the PCV was still low and the first reticulocytes appeared in the blood; 3 mo, 2 mo, 1 mo = serum drawn months after full hematologic remission. (From Dessypris et al., 1982.)

row cells was also investigated by Inoue et al. (1980). In their studies the peripheral blood mononuclear cells and purified T-lymphocytes from three children with TEC were shown to exert no inhibitory effect on autologous marrow CFU-E and/or BFU-E growth in vitro. At the present time it is unknown whether lymphocytes play any role in the suppression of erythropoiesis in children with TEC. These studies are relatively difficult to perform because they require a substantial amount of marrow with a high cell yield which will allow quantification, typing, and separation of bone marrow lymphocytes, as well as coculture studies of marrow lymphocytes with marrow lymphocyte–depleted cells.

A unique case of recurrent erythroblastopenia of childhood was also reported by Freedman (1983). The patient, a 9-month old previously normal child, had two episodes of prolonged anemia and erythroblastopenia within a period of 2 to 3 years. His marrow and peripheral blood mononuclear cells gave rise to normal numbers of CFU-E–and/or BFU-E–derived erythroid colonies. The patient's serum, drawn during each one of the episodes of erythroblastopenia, was found to be inhibitory for autologous blood BFU-E–derived erythroid colonies in vitro and this inhibitory activ-

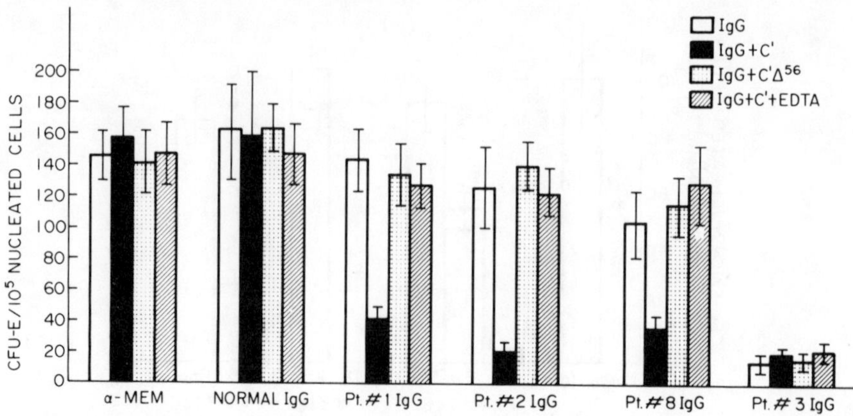

Figure 22 The effect of pretreatment of normal marrow cells with IgG purified from the sera of patients with transient erythroblastopenia of childhood, with or without complement, on the growth of normal marrow CFU-E in vitro. C′ denotes fresh human serum as a source of complement; $C'\Delta^{56}$ denotes human serum heat-treated at 56° C for 30 minutes. In patients 1, 2, and 8, incubation of marrow cells for 1 hour at 37° C with the patient's IgG and complement, followed by washing of the cells with medium, resulted in a significant loss of CFU-E, indicating a complement-dependent toxic effect of the IgG on the CFU-E itself. In patient 3, the IgG exerted a toxic effect on CFU-E through a complement-independent mechanism. The results represent the mean ± sem of quadruplicates from three different experiments. (From Dessypris et al., 1982.)

ity was localized to the IgM fraction of the serum. It should be noted, however, that in these studies not only the patient's but also normal serum IgM suppressed blood BFU-E growth in vitro by at least 50 percent. It was concluded that recurrent erythroblastopenia of childhood may be a variant of TEC that lasts longer, has the tendency to recur, is prednisone-responsive, and is mediated through a serum IgM and not IgG inhibitor. The existence of this entity of IgM-induced recurrent erythroblastopenia of childhood requires further confirmation.

A role for T-lymphocytes in mediating the inhibition of erythropoiesis has been suggested by the work of Hanada et al. (1985). They studied the effect of peripheral blood T-lymphocytes on the growth of autologous marrow CFU-E in a case of a 5-year-old girl with TEC. The addition of cryopreserved peripheral blood T-lymphocytes collected during the acute phase of the disease to autologous marrow cells at a ratio of 1:1 resulted in a 55 percent inhibition of CFU-E growth in vitro, whereas the addition of an equal number of blood T-lymphocytes collected after spontaneous remission of the erythroblastopenia had no effect on erythroid progenitor cell development. The magnitude of T cell–mediated inhibition of erythropoiesis was related to the number of added T-lymphocytes. Although these

Table 10 A Summary of the Results of in Vitro Studies in Twelve Children with Transient Erythroblastopenia of Childhood

	Erythroid progenitor cell growth in vitro		Effect of IgG on Autologous Marrow		Effect of IgG on Normal Marrow		
Patient No.	CFU-E	BFU-E	CFU-E	BFU-E	CFU-E	Preincubated CFU-E	BFU-E
1	N	N	↓	↓	↓	↓*	↓
2	N	N	↓	↓	—	↓*	—
3	N	N	↓	↓	↓	↓†	↓
4	D	N	PG	PG	—	—	—
5	D	D	PG	PG	↓	—	↓
6	D	D	PG	PG	↓	—	↓
7	N	N	—	—	—	—	—
8	ND	ND	ND	ND	—	↓*	—
9	ND	ND	ND	ND	—	—	↓
10	ND	ND	ND	ND	—	—	↓
11	N	N	—	—	—	—	—
12	D	D	PG	PG	—	—	—

Source: From Dessypris et al., 1982.
Note: N = normal; D = decreased; ND = not done; PG = poor growth; ↓ = inhibition; — = no effect
*Complement-dependent
†Complement-independent

studies suggested a possible role for T cells in mediating suppression of erythropoiesis, the fact that the T-lymphocytes were collected from the peripheral blood and not from the bone marrow, the absence of bone marrow lymphocytic infiltration, and the lack of data on the specificity of T cell action on erythroid and not granulocytic progenitors make the interpretation of the above findings difficult and raise questions about the significance of these findings for the pathogenesis of erythroid aplasia.

Congenital Hypoplastic Anemia (Diamond-Blackfan Syndrome)

Although the hereditary nature of this disorder is highly suggested by multiple reports in which the same hematologic abnormality was present in two or more successive generations (see chapter 1), it has not yet been established beyond a doubt, and the pattern of inheritance (autosomal dominant or recessive) seems to vary in different families. Alternatively, the congenital nature of CHA is supported by the fact that in 90 percent of cases there is only a single affected member in the family, and in 25 percent of affected children there is another coexisting congenital abnormality and an abnormal perinatal history. It is possible that what is recognized morphologically as CHA may be the result of various defects, some of which are hereditary and some of which are congenital.

The pathogenesis of CHA remains obscure, and the major defect(s) that leads to erythroid hypoplasia or aplasia is still unknown. Initial reports suggesting the presence of an abnormal metabolism of tryptophan in most affected infants—possibly responsible for the erythroid hypoplasia (Altman and Miller, 1953)—were not confirmed in subsequent studies (Gordon and Varadi, 1962; Diamond et al., 1976), and, furthermore, the described abnormalities were found to be nonspecific and present in other forms of anemia of diverse etiologies (Hankes et al., 1968; Price et al., 1970). An increased activity of orotate phosphoribosyltransferase and orotidine monophosphate decarboxylase in the red cells of five cases of prednisone-responsive CHA has been described (Zielke et al., 1979). Although these findings indicate that there is a basic metabolic alteration in the nucleic acid synthesis that is limited to erythroid cells in this disorder, how such an abnormality is linked to the basic defect responsible for the erythroid aplasia is not understood.

As in other cases of erythroid aplasia, the levels of erythropoietin in the sera of patients with this syndrome were found to be appropriately elevated for the degree of their anemia, so erythropoietin deficiency cannot be considered responsible for the severely decreased erythroid activity in the marrow of these patients (Hammond and Keighley, 1960; Gordon and Varadi, 1962; Falter and Robinson, 1972; Diamond et al., 1976; Alter, 1980).

In contrast to acquired PRCA in adults, CHA yields no conclusive evidence for the presence of humoral inhibitors of erythropoiesis or for lymphocyte-mediated suppression of erythroid cell development. Initial observations indicating the presence of inhibitors of ^{59}Fe-heme synthesis by marrow erythroid cells in vitro in the serum of patients with CHA (Ortega et al., 1975) were not confirmed in subsequent studies (Freedman et al., 1975; Geller et al., 1975). In the absence of humoral inhibitors of erythropoiesis in the sera of patients with CHA, a search for cell-mediated suppression of erythropoiesis in this disorder gave initially positive results. Peripheral blood lymphocytes from patients with CHA were reported to suppress normal marrow CFU-E growth in vitro (Hoffman et al., 1976; Steinberg et al., 1979); however, these findings were not confirmed or linked to the disease activity in many subsequent studies (Freedman and Saunders, 1978; Nathan et al., 1978*c*; Sawada et al., 1985). An increase in the number of T-lymphocytes with receptors for the gamma chain of the

IgG has been reported in the marrow and blood of a patient with CHA. The number of these cells returned to normal levels after prednisone treatment (Cornaglia-Ferraris et al., 1981). In view of the fact that T_γ cells suppress erythropoiesis in cases of chronic lymphocytic leukemia, further studies are warranted to investigate in an autologous system whether their presence in the marrow is linked with the pathogenesis of the anemia in this disorder.

Studies aiming at quantifying the number of erythroid progenitors in the marrow and blood in patients with CHA and the responsiveness of these erythroid cells to erythropoietin have provided some basic information regarding the nature of the defect in erythroid cell development in CHA. Freedman and co-workers (1976) studied four patients with CHA. Even in the absence of recognizable erythroblasts in these patients' marrows, marrow cells from all four patients gave rise to erythroblasts derived from the CFU-E. Two of the patients, while receiving no therapy, had decreased numbers of CFU-E in their marrows; the other two, while being treated with prednisone, had normal numbers of CFU-E. In this study it was also shown that marrow cells from these four CHA patients, although capable of giving rise to CFU-E–derived erythroblasts in vitro, compared to normal marrows had a significantly decreased number of CFU-E. Most important, a plateau of CHA marrow CFU-E growth in vitro was achieved with 2 to 2.5 times the concentration of erythropoietin that provided a maximum growth for normal marrow CFU-E.

These observations were further confirmed and extended by Nathan and co-workers (Nathan et al., 1978*b*; Nathan and Hillman, 1978). In their study of eight patients with CHA it was demonstrated that both marrow CFU-E and BFU-E, as well as blood BFU-E, were severely decreased in this disorder compared to normal controls (figure 23), and the responsiveness of these cells to erythropoietin in vitro was almost tenfold lower than that of normal cells (figure 24). Treatment with corticosteroids seems to restore the responsiveness of CFU-E to erythropoietin almost to normal levels. In the same studies it also became clear that the low erythroid colony formation in vitro in CHA affects much more severely the earliest erythroid cell assayable in vitro, the blood BFU-E, and less so the latest erythroid progenitor, the marrow CFU-E. It seems that erythropoiesis in CHA is arrested at a very early stage of erythroid development. Since early events in erythroid differentiation have been shown to be facilitated, at least in vitro, by products of T-lymphocytes and monocytes (Nathan et al., 1978*a*; Zuckerman, 1981), Nathan and co-workers (1978*c*) examined the erythropoietic helper T cell function in the blood of patients with CHA and found it to be normal when assayed on normal blood null cells which contain the erythroid progenitors. Therefore, T-lymphocytes in CHA can augment early erythroid development in vitro as efficiently as normal T cells, so an ab-

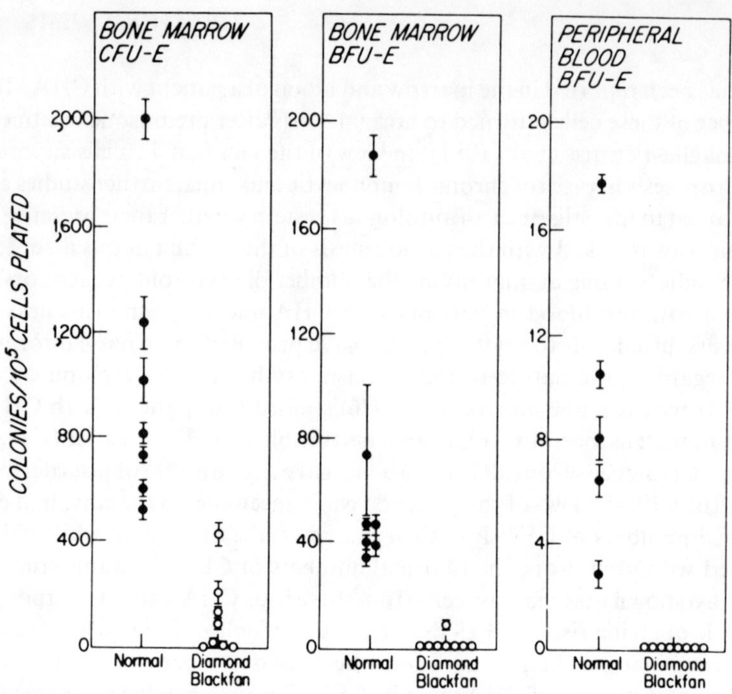

Figure 23 The proliferative capacity of bone marrow and blood erythroid progenitors in vitro in eight children with congenital hypoplastic anemia. (From Nathan and Hillman, 1978.)

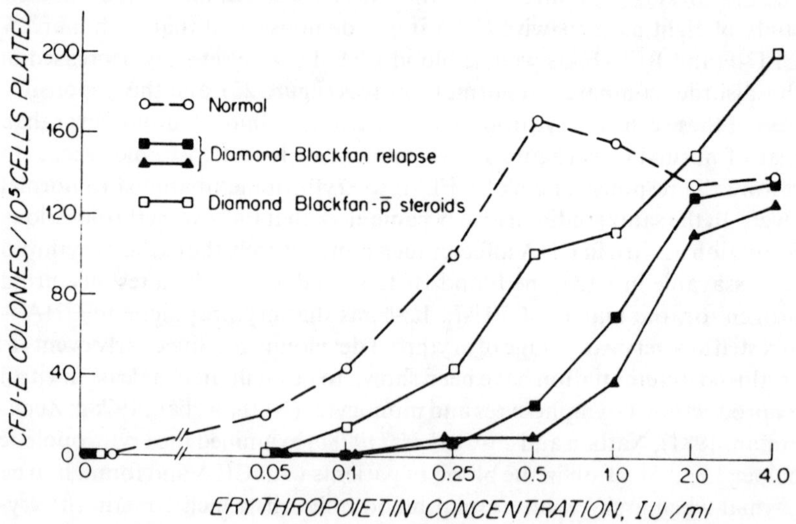

Figure 24 The response to erythropoietin in vitro of bone marrow CFU-E in congenital hypoplastic anemia. (From Nathan and Hillman, 1978.)

normal T cell erythropoietic growth–promoting function cannot be considered responsible for the arrest of erythropoiesis at such an early stage (Nathan et al., 1978*c*).

As has been pointed out, it is not known whether erythroid progenitors are indeed very low or absent from the marrow of CHA or if they are functionally abnormal so they could not give rise to erythroblastic colonies in vitro under the culture conditions. Whether the poor responsiveness of erythroid progenitors to erythropoietin is due to an intrinsic stem cell defect or to a microenvironmental factor, not yet identified, in the marrow of these patients is not known. Chan and colleagues (1982) speculated, based on the findings of marrow cell cultures from two infants with CHA, that this syndrome is pathogenetically heterogeneous with at least two recognizable forms, one in which there is an adequate number of erythroid progenitors that are relatively insensitive to erythropoietin, and a second form in which there is either a marked deficiency of erythroid progenitors or an absolute erythropoietin insensitivity. Restoration of sensitivity to erythropoietin during treatment with corticosteroids was also confirmed in their study in one of the two infants.

Another study of the erythroid progenitor cell differentiation in vitro in nine children with CHA produced almost the same results and led to the conclusion that CHA is not only the result of deficient formation of erythroid progenitors but is in addition a disorder due to defective erythroid differentiation (Lipton et al., 1986). It seems that with the currently available methodology for studying erythroid cell differentiation in vitro, it is very difficult to differentiate with any degree of certainty among low erythroid progenitor cell growth due to low numbers of erythroid progenitors, from low growth due to insensitivity of progenitors to erythropoietin, or from low growth due to an as yet unidentified suppressor cell that remains with the erythroid progenitors during the cell separation procedures.

Finlay and co-workers (1982) suggested, on the basis of a number of abnormalities of T cell function that they detected in five patients with CHA (three of whom had never received blood transfusions), that in CHA the basic defect is not limited to the erythroid cells but also involves the lymphocytes and the immune system in general. Ershler and colleagues (1980) presented evidence for a microenvironmental defect in the marrow of a patient with congenital hypoplastic anemia. They studied erythroid colony formation in the methycellulose system and observed that this patient's marrow, assayed twice, gave rise to normal or increased numbers of CFU-E despite the absence of erythroblasts in vivo. In further work they studied iron uptake and heme synthesis by either marrow cells in suspension or by marrow cells left in close contact with the bone in whole bone fragments. They found that there was a striking discrepancy in heme synthesis between marrow cells in suspension and marrow cells in the bone fragments, the latter showing no heme synthesis at all and markedly decreased iron uptake compared to the former. These findings suggest that, in this case at least, there was a bone marrow microenvironment that adversely affected erythroid differentiation. How frequently such a microenvironmental factor is responsible for the erythroid aplasia in CHA and the nature of such a factor are not yet known.

Conclusions and Unanswered Questions

On the basis of the presented studies it can be concluded that thus far there are at least five different mechanisms that have been identified and considered responsible for the induction of selective erythroid aplasia (see table 11).

Humoral inhibition of erythropoiesis by an IgG inhibitor has been found and studied in a significant number of cases. The in vitro and in vivo actions of this serum inhibitor are summarized in table 12. The target cell for the IgG inhibitor most frequently is an erythroid cell at a level of differentiation of CFU-E or between CFU-E and proerythroblasts. The nature of the antigen on these erythroid cells that is recognized by the inhibitory IgG is still unknown. Marmont (1977) proposed that PRCA may be an autoimmune receptor disease in the sense that the erythropoietin receptor on erythroid cells is the antigen recognized by the serum IgG inhibitor. This hypothesis is quite attractive because it provides an explanation for the selectivity of the erythroid aplasia and an understanding of the pathogenesis of the erythroid aplasia. However, recent work performed on mouse spleen cells infected with the anemia-producing strain of Friend virus (FVA cells) showed lack of interference of sixteen PRCA and eight TEC sera IgG with binding of radioactive erythropoietin to erythropoietin receptors present on FVA cells, despite the fact that eight of sixteen PRCA sera IgG and five of eight TEC sera IgG inhibited the erythropoietin-induced heme synthesis by the homogeneous population of FVA cells in vitro (Krantz and Goldwasser, 1987). Although these studies demonstrated that the IgG inhibitor of erythropoiesis present in the sera of a number of patients with PRCA or TEC does not have properties of an antibody to erythropoietin receptor they do not necessarily exclude this possibility, since the studies were performed on murine CFU-E–like cells and not on human CFU-E and the across-species reactivity of this antibody is not known. In addition, it is conceivable that the IgG binds to an epitope of the EP-receptor different from the one to which erythropoietin binds, so that degradation of the EP-receptor may occur without true competition of binding between EP and IgG. Until a method for preparation of homogeneous population of human CFU-E becomes available, and until the human EP receptor is isolated and characterized, negative results on the

Table 11 The Pathogenesis of Pure Red Cell Apasia

1. Humoral inhibition of erythropoiesis (IgG)
2. T-cell–mediated suppression of erythropoiesis
3. Parvovirus-induced toxicity to erythroid cells
4. Clonal stem cell defect (preleukemia)
5. Unknown mechanisms

Table 12 Actions of the Serum Inhibitor of Erythropoiesis in Pure Red Cell Aplasia

In vitro
Inhibition of ^{59}Fe-heme synthesis by marrow erythroblasts
Complement-mediated cytotoxicity to erythroblasts
Inhibition of marrow CFU-E and/or BFU-E growth and maturation
Complement-dependent or -independent cytotoxicity to CFU-E
Antierythropoietin antibody
In vivo
Suppression of ^{59}Fe-incorporation into newly produced red cells
Decrease of the pool size of marrow CFU-E

relation of PRCA IgG inhibitor to human EP-receptor should be interpreted with caution. It should be noted also that similar studies performed on heterogeneous bone marrow cell populations, or with the use of impure preparations of radioactive erythropoietin, have a great number of methodologic flaws and their results cannot be considered conclusive (Mladenovic et al., 1985).

A better understanding of the pathogenesis of PRCA, whether IgG-mediated, T-lymphocyte–mediated, or virus-induced, will await further progress in the basic field of erythroid cell differentiation.

4. Treatment and Natural History

Historically a variety of agents have been used for the treatment of acquired PRCA, including cobalt, riboflavin, pyridoxine, androgenic steroids, ACTH, corticosteroids, splenectomy, thymectomy for neoplastic or nonneoplastic thymus, cytotoxic immunosuppressive agents, antithymocyte serum or gamma globulin, high dosages of intravenous human immune gamma globulin, plasmapheresis, and cyclosporin.

Cobalt chloride has been used for its stimulatory effect on erythropoiesis through an increase in the levels of serum erythropoietin (Seaman and Kohler, 1953; Voyce, 1963). Subsequent trials of cobalt were without success (Donnelly, 1953; Sakol, 1954; Tsai and Levin, 1957). Its toxicity and the fact that erythropoietin levels are very elevated in this disorder make it an unreasonable and undesirable form of therapy. Riboflavin was reported to induce remission in a single case of PRCA with a history of exposure to mepacrine. Recovery was noted after 12 days of treatment (Foy and Kondi, 1953*a*). It is not known whether this patient had a mepacrine-induced case of PRCA in which erythropoiesis recovered after discontinuation of mepacrine, or whether the riboflavin antagonized the metabolic effects of this antimalarial drug, which is a riboflavin antagonist (DiGiacomo et al., 1966). Further trials of riboflavin in primary acquired PRCA have been unsuccessful (Tsai and Levin, 1957). The initial hypothesis, which later was proven to be correct, that a number of cases of PRCA have an autoimmune pathogenesis led to the wide use of corticosteroids alone (Gasser, 1951; Tsai and Levin, 1957; Schmid et al, 1963; Roland et al., 1964; DiGiacomo et al., 1966; Finkel et al., 1967; Kantz and Kao, 1969; Krantz 1972*b*, 1983; Beard et al., 1978) or in combination with cytotoxic immunosuppressive agents (Krantz and Kao, 1969; Böttiger and Rausing, 1972; Vilan et al., 1973; Bruntsch et al., 1974; Zucker et al., 1974; Beard et al., 1978; Peschle et al., 1978; Hunter et al., 1981; Krantz, 1983; Clark et al., 1984; Dessypris and Krantz, 1985*a*). These drugs constitute today the first treatment of choice.

The Initial Evaluation

After the diagnosis of PRCA is confirmed by bone marrow examination, all drugs that the patient receives should be discontinued and any infections present must be treated with the appropriate chemotherapeutic agents. If evidence exists for the presence of vitamin B_{12} or folate deficiency, these vitamins should be given in adequate doses. In the presence of an underlying nonthymic, nonhematopoietic malignancy appropriate therapy of the malignant neoplasm should be instituted. The presence of a thymoma must be sought by computerized tomography of the chest and if such a tumor is present thymectomy should be performed. During this period of initial evaluation red cell transfusions can be given as necessary.

In cases of drug-induced PRCA, the elimination of the drugs or treatment of infection is followed within a period of 1–2 weeks by the recovery of erythropoiesis. In cases of primary PRCA one may decide to wait for a period of 4–8 weeks before instituting specific treatment with the rationale that PRCA may run a short and self-limited course. If after such a waiting period no signs of recovery of erythropoiesis appear (reticulocytosis or stabilization of hematocrit), specific treatment must be instituted.

It should be noted that the morphologic diagnosis of PRCA based on the findings of bone marrow examination is sufficient to justify the initiation of therapy. Thus far no single laboratory test has been found to be able to confirm the diagnosis and predict the results of treatment. Generally patients whose marrow cells respond to erythropoietin in vitro, either by an increase of heme synthesis or by formation of erythroid colonies derived from CFU-E or BFU-E, have a higher chance of responding to some form of immunosuppressive therapy than do patients whose marrow cells do not respond in vitro to erythropoietin (Krantz, 1972*b*; Lacombe et al., 1984; Abkowitz et al., 1986). However, these in vitro studies are not sensitive enough to be used as criteria for therapeutic decisions.

Initial Immunosuppressive Therapy

In young adults and in patients without any contraindications for therapy with corticosteroids, prednisone should be given at a dose of 60–80 mg orally per day. This treatment should be continued for at least a period of 4 weeks. If a response is to occur, it usually appears within a mean period of 2.5 weeks (Clark et al., 1984). A rising reticulocyte count with a rising or stable hematocrit are the first hematologic signs of response to corticosteroids. After the hematocrit reaches a level equal to or higher than 35 percent, the dosage of prednisone is tapered slowly. If within 4–6 weeks after initiation of treatment no response has occurred the dosage of prednisone can be tapered rapidly.

In patients responding to prednisone, tapering of the dosage must be very gradual and slow, preferably over a period of 3–4 months at which time the prednisone can be finally discontinued. As with other autoimmune diseases, rapid tapering of prednisone may result in recurrence of PRCA. During this period of dosage reduction, close observation of the hematocrit and the reticulocyte count may provide some information as to whether the patient will require moderate or small doses of prednisone to be maintained in remission, or whether prednisone can be discontinued without the risk of immediate relapse. A number of patients seem to be steroid-dependent and in these patients the dose of prednisone that is needed to maintain a normal hematocrit varies widely. If small doses of prednisone are required (5–10 mg daily) the treatment can be continued for a period of 3–4 more months before the next attempt to further decrease the dose and eventually discontinue the drug. If higher doses, in excess of 10 or 15 mg daily, are necessary to maintain a normal hematologic picture, the addition of a cytotoxic agent should be considered.

An increased incidence of infections, retention of water and salt with exacerbation of preexisting hypertension or labile hypertension, hyperglycemia or exacerbation of preexisting diabetes, myopathy, and an increased incidence of peptic ulcer disease are the usual side effects from this short-term treatment with prednisone. Most patients are expected to increase their body weight and a number of them will develop a moon-face. Despite all these side effects prednisone alone is considered the first treatment of choice in young adults with PRCA mainly because of the lack of a leukemogenic and teratogenic effect. The response rate to prednisone alone is 45 percent for primary PRCA; however, no responses were noted in cases with secondary PRCA (Clark et al., 1984) (see table 13).

In patients past the childbearing age, patients with contraindications for high-dose prednisone treatment, and patients who have failed to respond to prednisone, the use of cytotoxic immunosuppressive agents, such as cyclophosphamide or azathioprine, alone or in combination with low doses of prednisone is recommended. In patients with an absolute contraindication for corticosteroid therapy either one of these two cytotoxic agents is given initially at a dose of 50 mg orally per day. In the absence of such a contraindication prednisone at a dose of 20–30 mg daily is added to the cytotoxic agent. The patient is followed weekly with complete blood counts and a reticulocyte count, and if the white blood and platelet count allow it, the dose of the cytotoxic drug is increased by 50 mg daily at weekly intervals up to a total daily dose of 150 mg. This dosage is maintained for a minimum of 3–4 months with weekly or biweekly checking of the white blood cell, platelet, and reticulocyte count. With this form of treatment the mean time to remission is 11 weeks with a range of 2–26 weeks (Clark et al.,

Table 13 The Results of Treatment of Pure Red Cell Aplasia at Vanderbilt-Affiliated Hospitals

	Number of Patients	
	Responding/Treated	Response Rate (%)
Antithymocyte globulin	2/6	33
Cyclosporin	3/4	75
Cytotoxic drugs		
with prednisone	23/41	56
without prednisone	1/13	8
Multiple treatments	35/49	72
Splenectomy	4/23	17
Prednisone alone	18/41	44
Spontaneous remission	5/49	10
No remission	14/49	28

Note: The total number of patients was 49. Of those, 32 had primary PRCA and 15 had secondary PRCA. Many patients suffered relapses and were treated with different regimens so one patient may be included in more than one treatment modality.

1984). If remission occurs within the first 12 weeks, the doses of cyclophosphamide or azathioprine and prednisone are slowly and progressively decreased and eventually both drugs are discontinued after 3–4 months from the time of normalization of hematocrit (see figure 25). In patients not responding to this regimen within 12 weeks of treatment an attempt should be made to increase the dose of the cytotoxic agent to the maximum tolerable one, which usually falls within the range of 200–250 mg daily while the dose of prednisone is maintained at 20–30 mg daily. With this increase one attempts to achieve the maximum of immunosuppression. This relatively high daily dose of cyclophosphamide or azathioprine is given under close supervision of white cell and platelet counts and until reticulocytosis or an increase or stable hematocrit is seen, or the absolute granulocyte count falls below 1,500/μl, or the platelet count drops below 90,000/μl. In case of response the dose is decreased slowly and both the cytotoxic agent and prednisone are eventually discontinued as described above. If granulocytopenia or thrombocytopenia develops, the cytotoxic agent is discontinued while the prednisone is maintained. During the period of marrow recovery with the return of granulocyte and platelet count to normal, a response is noted in a number of patients as indicated by an increase of the reticulocyte count and stabilization of their hematocrit (Krantz et al., 1978; Hunter et al., 1981; Clark et al., 1984) (figure 26). In the case of such a response, the dosage of the cytotoxic agent and prednisone are slowly tapered and finally discontinued in a fashion already described. If no response occurs, this

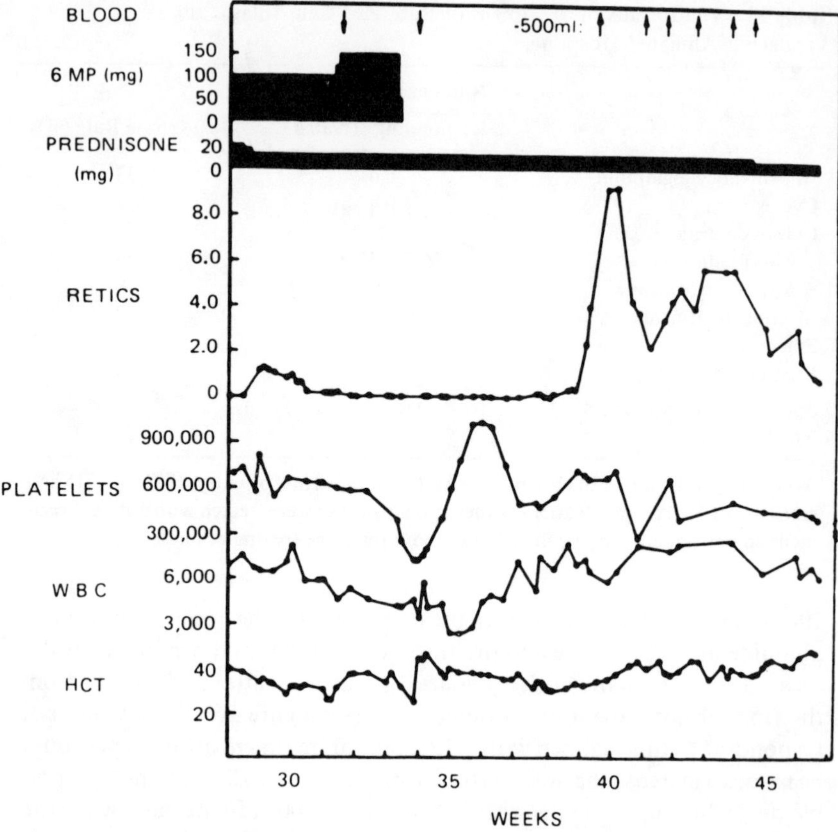

Figure 25 The early response to prednisone and 6-mercaptopurine in a patient with pure red cell aplasia. (From Krantz and Kao, 1969.)

regimen should be considered a failure, the cytotoxic agent and the prednisone should be discontinued, and the patient maintained on red cell transfusions as needed.

Although the response rate of both primary and secondary PRCA to the combined treatment with prednisone and a cytotoxic agent seems to be superior to that with prednisone alone (56 versus 37 percent according to Clark et al., 1984; and in the updated Vanderbilt series 56 versus 44 percent) the degree of immunosuppression is more pronounced and the incidence of infections may be higher. In addition to myelosuppression, hemorrhagic cystitis and hepatotoxicity constitute the two most frequent side effects from cyclophosphamide and azathioprine therapy, respectively.

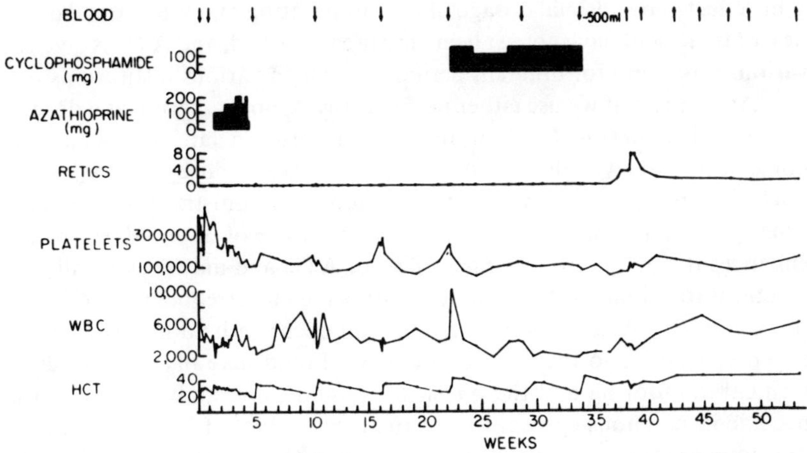

Figure 26 The response to cyclophosphamide (Cy) and prednisone appearing after mild marrow toxicity induced by cyclophosphamide in a patient with pure red cell aplasia. (From Krantz, 1978.)

The Management of Nonresponders

Patients not responding to either prednisone or cytotoxic immunosuppressive agents, alone or in combination, should be considered for one of the following forms of treatment.

Antithymocyte Gamma Globulin

Antithymocyte gamma globulin (ATG) is prepared from the serum of horses or rabbits that have been immunized with human T-lymphocytes usually collected by cannulation of the thoracic duct. Although this preparation has a high titer of antihuman T cell antibodies, it also contains antibodies against other human cells and at least in vitro in high concentrations is toxic for a variety of human cells including hematopoietic cells (Greco et al., 1983; Dessypris and Krantz, unpublished observations).

The action of ATG is not limited only to selective suppression of cell-mediated immunity, and a decrease of peripheral blood lymphocytes is noted only temporarily during its administration. This form of treatment has been used successfully in a number of patients with PRCA (Krantz, 1972*a*; Vilan et al., 1973; Marmont et al., 1975; Peschle et al., 1978; Hagberg et al., 1980; Lacombe et al., 1984; Marmont, 1984; Harris and Weinberg, 1985; Jacobs et al., 1985; Marmont et al., 1985; Abkowitz et al., 1986). It has the advantage that it lacks myelotoxicity and has no leukemogenic effect. The optimal dosage of this preparation and the optimal duration of treatment have not yet been firmly established, and ATG is given at various doses and for different periods of time in various institutions.

At Vanderbilt we use either rabbit antilymphocyte serum at a dose of 0.2–0.3 ml/kg in slow 4–6 hour intravenous infusion daily for 14 days, or horse antithymocyte globulin (Atgam, Upjohn) at a dose of 20 mg/kg of horse IgG per kg of body weight for 7 days. Administration of ATG is usually combined with oral prednisone at a dose of 30 mg daily, which enhances the immunosuppressive effect of ATG and may prevent allergic reactions to animal proteins that constitute the most frequent side effect of this treatment. Anaphylactic reactions can occur with administrations of any preparation, so the infusion of ATG should take place under close medical monitoring and the physician should have ready for use at the bedside intravenous epinephrine to treat them. More frequent than anaphylactic reactions are fever, skin rash, or manifestations of serum sickness that can be relatively easily managed by increasing the dose of the concomitantly administered steroids.

The effectiveness of ATG in the treatment of PRCA is not exactly known since the literature contains only reports of its successful use but not the failures. In a study of nine patients with PRCA reported by Abkowitz et al. (1986*b*) a response rate of 66 percent was observed. It should be noted, however, that four of the six responders had primary PRCA, one PRCA secondary to B cell CLL, and another one PRCA secondary to chronic infectious mononucleosis. None of the remaining patients with severe erythroid aplasia associated with a myelodysplastic or myeloproliferative syndrome responded to this treatment. ATG-induced remissions may be permanent or transient and ATG has been used in the same patients repeatedly with success many times for the treatment of recurrent PRCA. Studies comparing ATG to conventional immunosuppressive treatment are not available. Previous failure with steroids and cytotoxic agents does not seem to decrease the chances of responding to ATG.

Today ATG can be considered the treatment of choice in patients with primary or CLL-associated PRCA unresponsive to steroids, and a trial with ATG should be attempted before initiation of treatment with cytostatic agents, particularly in children, young patients, or women of childbearing age in whom the leukemogenic and teratogenic effect of cytostatic drugs must be avoided whenever possible.

Plasmapheresis

The demonstration of an IgG inhibitor of erythropoiesis in the serum of a number of patients with PRCA has led to the rational attempt to induce remission by removal of this IgG through plasmapheresis. Plasmapheresis has been used in a small number of patients with PRCA refractory to immunosuppressive therapy (Marinone et al., 1981; Messner et al., 1981; Lacombe et al., 1984; Freund et al., 1985; Khelif et al., 1985; Berlin and Lieden, 1986), and good results have been reported in at least two cases (figure 27). We have also seen in consultation a case of refractory PRCA and hemolytic anemia in which prolonged and intensive plasmapheresis resulted in recovery of erythropoiesis. Since IgG is distributed both in the intravascular and extravascular space, one cannot expect any dramatic results unless this procedure is repeated thrice weekly for at least 2–3 weeks. An effect from plasmapheresis can be detected by bone marrow reexamination before any reticulocytosis appears in the peripheral blood (Marinone et al., 1981). Today plasmapheresis can be considered the treatment of choice in cases refractory to steroids, cytotoxic agents, and ATG in which there is a contraindication to performing splenectomy.

Splenectomy

Although the spleen does not play any primary role in the destruction of marrow erythroid cells, nevertheless, it is a major B cell lymphoid organ with significant ability for antibody production. Initial reports on the effect of splenectomy in PRCA have provided negative results (Jacobs et al., 1959; Schmid et al., 1965; Hirst and Robertson, 1967; Krantz and Kao, 1969). Subsequent reports, however, have shown that in a number of patients refractory to conventional immunosuppressive treatment splenectomy may be followed by return of erythropoiesis to normal (Loeb et al., 1953; Chalmers and Boheimer, 1954; Eisemann and Dameshek, 1954; Zaentz et al., 1975; Beard et al., 1978; Krantz, 1983; Clark et al., 1984). Resumption of erythropoiesis may be observed in the first few weeks post splenectomy or later within the first two months after surgery. Patients not showing any evidence of recovering erythropoiesis within 8–10 weeks following splenectomy should be retreated with immunosuppressive drugs following the same approach as for initial therapy of PRCA. In a number of cases unresponsive to initial immunosuppression and splenectomy, a repeat course of immunosuppressive treatment has led to recovery of erythropoiesis (Dreyfus et al., 1963; Safdar et al., 1970; Vilan et al., 1973; Zucker et al., 1974; Krantz and Zaentz, 1977; Clark et al., 1984).

Splenectomy also should be considered in young patients with frequent relapses of PRCA that require chronic treatment with prednisone or cytotoxic agents. In one of our patients with a history of eight relapses within a period of ten years, who could not tolerate chronic immunosuppressive therapy, splenectomy allowed discontinuation of all maintenance therapy with normal hematologic values for more than two years following removal of the spleen. It seems that although splenectomy may not have a place in the primary management of PRCA, it may be of benefit in refractory or frequently relapsing cases of this disorder. One should keep in mind

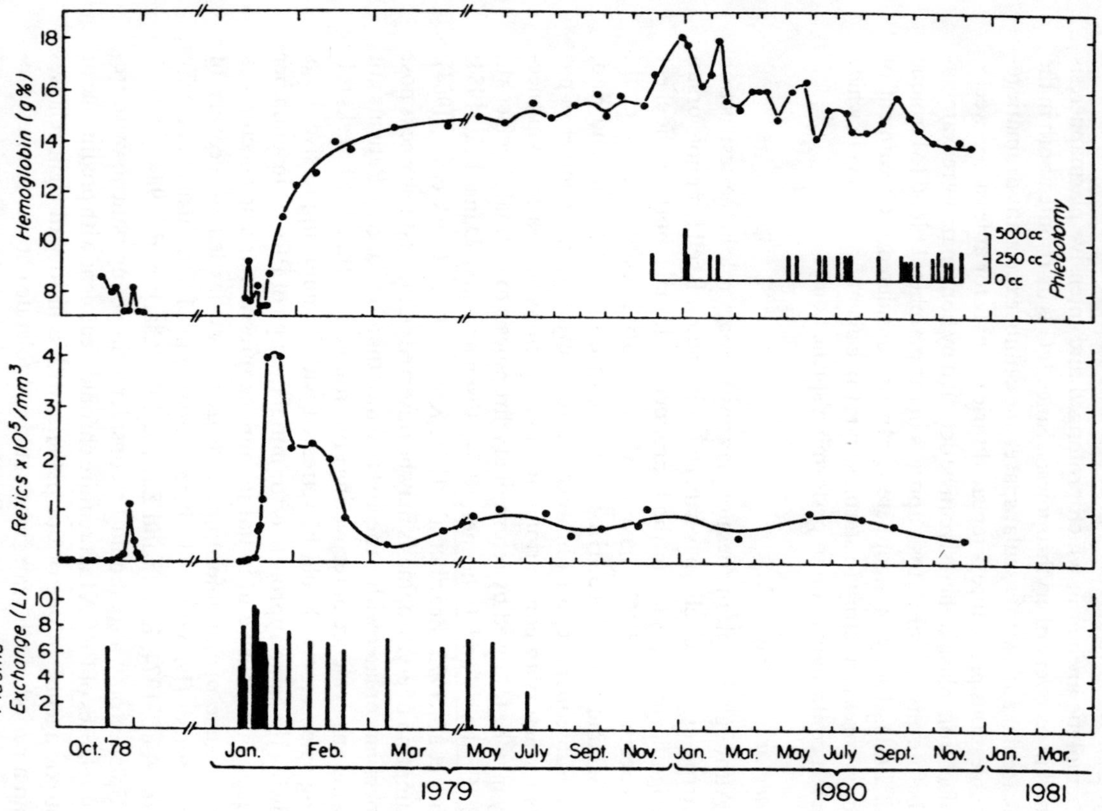

Figure 27 The response of a patient with pure red cell aplasia to plasmapheresis. (From Messner et al., 1981.)

that besides the surgical morbidity and mortality associated with this procedure, splenectomized patients have an increased incidence of severe, and frequently fatal, infections by encapsulated bacteria, most often pneumococci. Therefore, candidates for splenectomy must be taken off all immunosuppressive drugs and be vaccinated appropriately with polyvalent antipneumococcal vaccine. They must be made aware of this risk and advised to seek medical attention at the very beginning of any febrile illness.

Other Forms of Treatment

In patients refractory to all the above therapeutic modalities, the following forms of treatment also must be considered.

ANDROGENIC STEROIDS

Androgenic steroids have been shown to stimulate erythropoiesis by increasing the levels of erythropoietin and by acting on the early erythroid progenitor cells (Necheles, 1971; Madder et al., 1978). In a number of cases, androgenic steroids have led to erythropoietic recovery (Schmid et al., 1965; DiGiacomo et al., 1966; Hirst and Robertson, 1967). In other studies, however, androgenic steroids have shown no effect (Vilan et al., 1973; Krantz and Zaentz, 1977; Clark et al., 1984). In patients refractory to all forms of treatment, androgenic steroids can be used on a trial basis.

The type of the preparation to be given depends on the physician rather than on any firm evidence that one of them has any better erythropoietic activity than others. Our experience with androgenic steroids has been disappointing. These steroids should be given for at least three months before any decision is made regarding their effectiveness. Side effects include retention of water and salt, hirsutism and other signs of virilization in females, cholestatic jaundice with or without a hepatitislike picture, and, when used for prolonged periods of time, peliosis hepatis and development of hepatic adenomas. Recently, danazol, an attenuated synthetic androgenic steroid used previously for treatment of a number of autoimmune hematologic disorders, has been reported to have induced remission of PRCA in patients unresponsive to corticosteroids and cytostatic agents (Lippman et al., 1986). Since the number of reported cases is limited, the role of this agent in the treatment of PRCA remains to be determined.

HIGH-DOSE INTRAVENOUS GAMMA GLOBULIN

High-dose intravenous gamma globulin has been used initially for the treatment of immune thrombocytopenic purpura and autoimmune neutropenia (Imbach et al., 1981; Fehr et al., 1982; Pollack et al., 1982). The mechanism through which high doses of intravenous gamma globulin are capable of improving immune peripheral cytopenias is not exactly understood (Bussel and Hilgartner, 1984; Dwyer, 1987). A blockade of the Fc receptors of the macrophages in the reticuloendothelial system (Fehr et al., 1982), an induction of hemolysis of Rh^+ erythrocytes by the small amounts of anti-Rh IgG present in these preparations with subsequent decrease of the function of the splenic macrophages (Salama et al., 1983), and the infusion of antiidiotypic antibodies capable of suppressing synthesis of specific autoantibodies (Sultan et al., 1984) have been proposed.

Regardless of the exact mechanism through which high doses of gamma globulin improve immune cytopenias, it seems that pharmacologic concentrations of gamma globulin exert significant immunomodulatory effects, thus explaining the induction of prolonged unmaintained remissions of these disorders following a single course of intravenous gamma globulin. In recent studies such a treatment resulted in a significant decrease of all immunoglobulin production in vitro (78 percent decrease of the IgG synthesis) and in an increase in the number of circulating suppressor T-lymphocytes (Tsubanio et al., 1983; Delfraissy et al., 1985).

Clauvel et al. (1983) reported their experience from the use of high-dose gamma globulin in four patients with PRCA; one idiopathic, one associated with a well-differentiated lymphocytic lymphoma without marrow involvement, and two cases of PRCA associated with B cell CLL. The infusion of 0.4 g/kg/d for 5 consecutive days resulted in recovery of erythropoiesis in the two cases associated with CLL, which became apparent by an increase of reticulocyte count as early as day 5–8 of treatment. One of these patients relapsed 3 weeks later and was successfully retreated with the same regimen. Based on their findings of marrow examination in ITP patients responding to high-dose gamma globulin, these investigators suggested that high-dose gamma globulin may have an effect through improving hematopoiesis in the marrow in various cytopenic states. The beneficial effect of high-dose gamma globulin in PRCA has been confirmed in subsequent reports (Etzioni et al., 1986; Katakkar, 1986; McGuire et al., 1986). Recently at Vanderbilt Medical Center this form of treatment was successfully used in two cases of PRCA associated with B cell CLL. The role of high-dose gamma globulin in the management of PRCA is difficult to determine at this point in time and further studies of its indications are necessary. Considering the extremely low incidence of side effects, however, this form of therapy must be thought of as an option in cases refractory to corticosteroids, especially in children and young adults, before the initiation of cytotoxic chemotherapy or consideration of splenectomy.

CYCLOSPORIN

Cyclosporin-A is a new immunosuppressive agent that has been primarily used for prevention of graft rejection after organ transplantation (Cohen et al., 1984). This agent also has been used for the treatment of various diseases of autoimmune pathogenesis (Nussenblatt et al., 1983; Stiller et al., 1984). Although a variety of effects of cyclosporin on the immune system have been described, including effects on T- and B-lymphocytes, the exact sequence of events that leads to immunologic tolerance and/or suppression of autoimmune phenomena has not yet been fully elucidated (Shevach, 1985).

Four patients with PRCA have been thus far reported to have responded to treatment with cyclosporin-A (Tötterman et al., 1984; Debusscher et al., 1985; Chikkappa et al., 1986*a*). We have recently used cyclosporin for the treatment of four patients with acquired primary PRCA refractory to all other immunosuppressive therapies including ALG and plasmapheresis. Cyclosporin was administered orally at a dosage of 10–12 mg/kg daily in two divided doses along with 30 mg prednisone a day. In three cases within 1 to 2 weeks a response was noted as indicated by a rising reticulocyte count and an increasing hematocrit, which reached normal levels by the second to third week of treatment. In the first patient the dosage was rapidly tapered and the drug discontinued after normalization of the hematocrit resulting in prompt relapse of the disease. Remission was again easily induced by a second course of cyclosporin.

Based on the currently available experience, it seems that patients responsive to cyclosporin after normalization of their hematocrit can be maintained in remission with very small dosages of this agent. Since nephrotoxicity constitutes the major and limiting side effect of cyclosporin treatment, progressive decrease of the dosage to the minimum one required for maintenance of remission seems to be wise. Treatment of PRCA with cyclosporin appears to be very promising, but such a treatment should be considered at this time still experimental. Further studies are necessary to determine the effectiveness of this drug, the optimal and least toxic dosage, the minimum duration of therapy for induction of remission, and whether or not there is need for maintenance treatment.

The Results of Treatment

Evaluation of the results of treatment of PRCA is extremely difficult because the disease has an unpredictable clinical course with spontaneous remissions occurring in 10–12 percent of patients within a period ranging from 4 months to 14 years after the diagnosis. In addition there is no clinical or laboratory criterion that can safely distinguish between chronic and acute PRCA. A number of cases that are treated successfully immediately after confirmation of diagnosis may in fact represent examples of acute PRCA secondary to viral infections or drugs that the patient did not mention. Comparing one therapeutic approach to another is almost impossi-

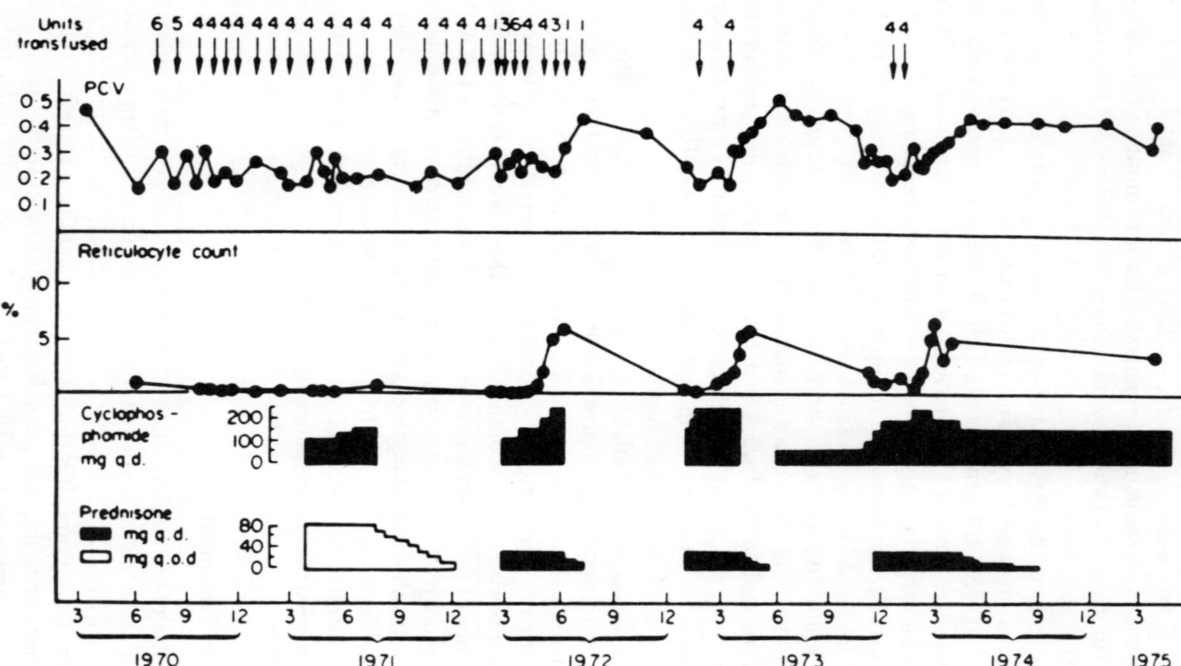

Figure 28 Repeated responses to cyclophosphamide and prednisone in a patient with relapsing pure red cell aplasia. (From Zaentz et al., 1976.)

ble, since the disease is rare enough that controlled studies are practically impossible to perform.

At Vanderbilt University–affiliated hospitals, 49 patients have been treated by various, and frequently multiple, forms of therapy; the results of their treatment are presented in table 12. A remission of the disease can be achieved in 72 percent of cases using a variety of immunosuppressive agents (Clark et al., 1984; and updated data from the Vanderbilt experience). Similar results have been reported by other investigators using slightly different immunosuppressive regimens. In a series of 21 patients with chronic PRCA reported by Peschle et al. (1978), 17 patients were treated with cyclophosphamide or ATG, alone or in combination, and a response was noted in 9 with an overall response rate of 53 percent. In a more recent report by Lacombe and colleagues (1984), 61 percent of 18 patients with acquired PRCA responded to some form of immunosuppressive treatment including corticosteroids, cyclophosphamide, antithymocyte globulin, azathioprine, or plasmapheresis. The recent addition of high-dose gamma globulin and cyclosporin to the spectrum of therapeutic modalities useful in PRCA and the continuously improving experience from the use of ATG in aplastic anemia and PRCA are expected to further improve the response rate. With long-term immunosuppressive therapy, however, it is reasonable to expect that almost 40 percent of treated patients will develop a form of infection (Clark et al., 1984).

Relapses and Maintenance Therapy

Although induction of a remission can be easily achieved in the majority of patients with one or the other immunosuppressive regimen, the remission may not be permanent and the majority of patients relapse following discontinuation of treatment. In the series presented by Clark et al. (1984), 13 of 23 patients who responded to treatment and 1 of 5 patients with spontaneous remissions have relapsed within the follow-up period. A life-table analysis of these data indicated that 80 percent of patients with induced remissions will relapse within 24 months following resumption of normal erythropoiesis. Treatment of relapses was almost equally successful with 10 of 13 patients entering a second or third remission (see figure 28).

The fact that a number of patients with PRCA require maintenance therapy to maintain normal erythropoiesis has been recognized in the past (Finkel et al., 1967; Zaentz et al., 1976), and maintenance therapy has been proposed and found to be successful (Zaentz et al., 1976; Krantz, 1983; Dessypris et al., 1984). Since the natural course of the disease in a single patient cannot be predicted by any clinical or laboratory parameter, the decision to initiate maintenance therapy is purely empirical. Maintenance therapy can be justified in patients with more than one relapse in a relatively short period of time. Splenectomy should be considered in these cases before any cytotoxic maintenance therapy is started. Alternatively, one may argue that since 80 percent of patients are going to relapse within 24 months following remission, maintenance chemotherapy should be given to all patients. This approach, although reasonable, may expose a greater number of patients to the long-term side effects of corticosteroids, or the leukemogenic effect of cytotoxic agents. If maintenance therapy is thought to be necessary, it can include either low-dose prednisone or cyclophosphamide, 10–20 mg or 50–100 mg daily, respectively. Maintenance chemotherapy once started should be given for at least 24 months unless severe side effects necessitate its premature discontinuation. Usually patients with a tendency to relapse remain free of relapses following a 1–2 year course of maintenance therapy. A summary of the main forms of remission induction and maintenance immunosuppressive treatment is presented in table 14. Although the order by which the different therapeutic approaches are presented in this table corresponds to the classic way of managing patients with PRCA, one should keep in mind that this order may change in the near future with acquisition of greater experience with ALG and cyclosporin, and that other factors, such as age or coexisting diseases, should be taken into account when deciding which therapeutic modality will be used.

Thymomas and Pure Red Cell Aplasia

In patients presenting with PRCA associated with thymoma, thymectomy should be performed before any immunosuppressive therapy is considered. Within 4–8 weeks following thymectomy remission of erythroblastopenia occurs in 25–30 percent of cases (Jacobs et al., 1959; Roland, 1964; Hirst and Robertson, 1967). In a number of patients treated with steroids prior to thymectomy, removal of the thymoma improved the effectiveness of corticosteroids in inducing remission of PRCA (Hirst and Robertson, 1967). In the remaining 70 percent of patients in whom PRCA persists after thymectomy, immunosuppressive therapy should be given as already discussed for the primary form of this syndrome. Depending on the operative stage and the invasive or not nature of the thymoma other forms of treatment may also be indicated, such as radiotherapy or chemotherapy (Penn and Hope-Stone, 1972; Boston, 1976; Bergh et al., 1978; Wilkins and Castleman, 1979; Evans et al., 1980; Daugaard et al., 1983). Patients with thymoma and PRCA in whom there are contraindications for surgery can be considered candidates for radiotherapy. The effectiveness of radiotherapy alone in inducing remission of PRCA is not well known and both successes and failures have been reported (Roland, 1964; Field et al, 1968; Souquet et al., 1970; Penn and Hope-Stone, 1972).

Table 14 Immunosuppressive Treatment of Pure Red Cell Aplasia

Remission induction	
Prednisone	60 mg daily until remission or for 6–8 weeks
Cytotoxic drugs (cyclophosphamide or azathioprine)	2–3 mg/kg daily, titrated to maintain normal leukocyte and platelet count, until remission or for 3–4 months; if no remission occurs, increase dose until granulocyte count falls below 1500/μl or platelet count below 90,000/μl, then discontinue
Cytotoxic drugs and prednisone	Cytotoxic drugs as above plus prednisone 20–30 mg daily
Antihuman thymocyte globulin	10–40 mg of IgG/kg daily for 4–14 days with or without prednisone 30–60 mg daily starting the day before the first infusion and tapered rapidly over 1–2 weeks after completion of treatment
Splenectomy	
Plasmapheresis	3 exchanges weekly for 3–5 weeks (25–35 liters exchanged)
High-dose gammaglobulin	0.4 g/kg daily intravenously for 5 days
Cyclosporin-A	10–12 mg/kg daily, orally in 2 divided doses with 20–30 mg oral prednisone daily until the hematocrit normalizes, then rapid tapering of prednisone followed by slow tapering of the dosage of cyclosporin-A
Remission maintenance	
Cytotoxic drugs	1–2 mg/kg daily for 6–24 months
Prednisone	10–20 mg daily for 6–24 months

Removal of a normal thymus gland as a primary or secondary form of immunosuppressive treatment has been tried in a limited number of cases with unfavorable results (Soulter et al., 1957; Soulter and Emerson, 1960; Zeok et al., 1979; Krantz, 1986), and thymectomy in the absence of thymoma cannot be recommended.

Evolution into Other Hematologic Disorders

A number of cases have been reported in which PRCA evolved either into aplastic anemia or acute nonlymphocytic leukemia. Evolution of PRCA into global marrow aplasia with peripheral blood pancytopenia (aplastic anemia) can occur within a period of a few months to a few years following the initial diagnosis of PRCA. The development of peripheral cytopenia other than anemia in a patient with chronic PRCA should always be investigated by repeating the bone marrow aspiration and biopsy, since a similar peripheral blood picture is not infrequent in chronic, refractory to treatment, and transfusion-dependent cases of PRCA in which, following years of therapy with red cell transfusions, splenomegaly develops with hypersplenism and mild to severe pancytopenia. This is usually first manifested by an increase in the transfusion requirements. Development of aplastic anemia has been described both in cases refractory to all forms of therapy, as well as in cases following remission of the initial episode of PRCA. Aplastic anemia has developed as well in cases of idiopathic PRCA and in cases associated with thymoma (Schmid et al., 1963; Hirst and Robertson, 1967; Negri-Gualdi, 1977; MacMahon and Egan, 1980; Clark et al., 1984). The exact incidence of evolution of PRCA into aplastic anemia is not known.

Acute nonlymphocytic leukemia is another disorder into which PRCA may eventually develop (Schmid et al., 1963; Bernard, 1969; Pierre, 1974; Savage, 1976; Peschle et al., 1978; Kato et al., 1979; Moloney, 1979; Clark et al., 1984). The frequency of leukemic transformation among patients with PRCA is difficult to assess, since PRCA is frequently confused with cases of myelodysplastic syndrome associated with erythroid hypoplasia (refractory anemia without excess of blasts). Schmid and colleagues (1963) reported 4 cases of acute leukemia among 39 cases of chronic primary PRCA but not among 37 patients with PRCA associated with thymoma. Pierre (1974), in a review of the literature on preleukemia, found 7 cases of PRCA in which acute leukemia resulted in death; however, the total number of cases of PRCA from which this number was collected is unknown. In a series of 22 patients with PRCA reported by Peschle and associates (1978), 4 developed acute leukemia. Among 47 cases of PRCA seen at Vanderbilt-affiliated hospitals, acute leukemia developed in 2. Based on these reports, it seems that acute leukemia develops in a proportion of 4–20 percent of patients with primary PRCA. The relation of PRCA to acute leukemia is poorly understood. In some cases selective erythroid aplasia may be the first manifestation of a preleukemic syndrome, and in other cases acute leukemia develops shortly after remission of the first episode of PRCA (Krantz and Dessypris, unpublished observations). It seems that primary acquired PRCA has a relatively lower frequency of transformation into acute leukemia than do the myelodysplastic syndromes (Dessypris et al., 1980).

Iron-Chelation Therapy for Patients Refractory to Treatment

Patients refractory to all forms of treatment, constituting about 30–40 percent of all patients with PRCA, are managed by red cell transfusions at regular intervals. The average red cell transfusion requirements for such patients range between 3–4 units of packed red cells for 4 weeks, depending on the patient's age, coexisting diseases, and the level of hematocrit one decides to maintain. It is, therefore, predictable that after a period of time averaging 2–3 years the total body iron stores will exceed 20–30 g and organ dysfunction from parenchymal iron deposition is expected to appear. In older patients with coexisting heart, liver, or pancreas disease the symptoms may be seen much earlier than in younger patients with the same degree of iron overload. Since a case of PRCA cannot be considered refractory to treatment until all therapeutic modalities have been tried without success, and since such trials cannot be completed in a time span shorter than 18–24 months, consideration of iron chelation treatment should be given in all refractory cases.

Chronic iron-chelation therapy consists of the subcutaneous infusion of 1.5 g to 2 g of deferoxamine during a 12-hour period daily by the use of a portable pump (Propper et al., 1976, 1977). Usually patients administer the drug to themselves from 6:00 to 7:00 P.M. to 6:00 or 7:00 A.M. of the following day, a schedule that does not interfere with their daily activities. The amount of iron secreted in the urine and stools depends on the total body iron. The excretion of iron chelated by deferoxamine can be further increased by combined administration of ascorbic acid (O'Brien, 1974). However, since administration of ascorbic acid has been associated with exacerbation of cardiomyopathy, and particularly the induction of cardiac arrhythmias (Nienhius, 1981), it is wiser to start oral ascorbic acid therapy after a period of iron-chelation therapy (1–2 months) and to prescribe it initially in small doses, which can later be increased provided no cardiac side effects have appeared. The effectiveness of iron-chelation therapy in patients dependent on periodic red cell transfusions has been well proven (Hussain et al., 1976; Propper et al., 1976, 1977; Weiner et al., 1978; Schafer et al., 1985). However, the effect of such a treatment on the overall survival of these patients is not yet known (Schafer et al., 1985).

Transient Erythroblastopenia of Childhood

Since in transient erythroblastopenia of childhood the arrest of erythropoiesis is transient, no specific treatment is needed. In the majority of cases complete recovery is seen within 4–6 weeks following the diagnosis. Red cell transfusions may be needed in a number of children with severe anemia and symptoms. The routine use of corticosteroids is not recommended. In contrast to acquired PRCA in adults, TEC does not recur. Only one case of recurrence of TEC has been reported in a child after another viral illness (Lovric, 1970).

Congenital Hypoplastic Anemia

Corticosteroids (prednisone) constitute the primary form of treatment today for congenital hypoplastic anemia. Prednisone is given at a dose of 2 mg/kg/day until the patient's hemoglobin concentration reaches 9–10 g/dl and then the dose is slowly decreased to the lowest one that can maintain the hemoglobin at a level between 8 and 10 g/dl. A response to prednisone is seen by an increase of the reticulocyte count within an average period of 1–2 weeks, which is later followed by an increase in the hemoglobin concentration. Once the lower effective dose of prednisone is found, to avoid the deleterious effects of corticosteroids on growth, the drug is given on an alternate-day schedule (Alter, 1980) or treatment is given daily for one week followed by two to three weeks off treatment (Sjölin and Wranne, 1970). The response to steroids may be complete and the remission can be maintained by a chronic maintenance low-dose therapy, or the response may be intermittent. Dependence on steroids for maintenance of remission is seen in the majority of patients. The responsiveness to steroids seems to vary at different times. Failure to respond to this dose of prednisone should be followed by a trial of higher doses (4 mg/kg/daily) or a combination of corticosteroids and androgenic steroids, or a trial of prednisolone or dexamethasone (Alter, 1980). Children taking drugs interfering with the metabolism of prednisone (diphenylhydantoin, phenobarbital) may respond only after discontinuation of such agents.

Despite the initial suggestion that response to steroids occurs more frequently the earlier such a treatment is initiated (Diamond et al., 1961), it seems that the duration of the disease does not necessarily affect the responsiveness to steroids (Schorr et al., 1960; Alter, 1980). Androgenic steroids by themselves are not proven to be effective in the treatment of this disorder (Diamond and Blackfan, 1938), and a number of cases that responded to androgenic steroids did so only in combination with corticosteroids (Tartaglia et al., 1966; Starling and Fernbach, 1967; Diamond et al., 1976). Splenectomy is indicated only in cases with increasing red cell transfusion requirements and a proven shortened red cell survival (Miller, 1978). Treatment with antithymocyte globulin has been tried without success in at least two patients with CHA (Haas et al., 1985). Recently, bone marrow transplantation has been used as a form of therapy in cases unresponsive to conventional treatment with corticosteroids. Bone marrow transplantation restored erythropoiesis to normal in both cases (August et al., 1976; Iriondo et al., 1984). The role of bone marrow transplantation in such refractory-to-treatment cases needs further study before any specific indications can be established.

Recently isolated cases of responses to high doses of intravenous methylprednisolone (Özsoyln, 1984) and to cyclosporin-A (Tötterman et al.,

1984) have been reported. These forms of therapy deserve further investigation to determine their role and effectiveness in the management of this anemia.

Evaluation of treatment results in CHA should take into account that spontaneous remissions do occur in this disorder at any time in the course of the disease. Among 213 cases reviewed by Alter (1980), the incidence of spontaneous remissions was as high as 25 percent and they occurred within 3 months to 16 years following the diagnosis.

The prognosis of this disorder improved significantly after the introduction of corticosteroid therapy. Almost 60–80 percent of treated cases are expected to respond to one or another form of such treatment. Cases refractory to steroids are managed by red cell transfusions and iron-chelation therapy. The mortality from this disorder is about 15 percent in reported cases. A significant number of deaths in the early reported cases were due to transfusional hemosiderosis (almost 25 percent), which can be prevented today with the use of deferoxamine. Infections constitute the major cause of mortality in this disease and almost 30 percent of deaths among the cases reviewed by Alter (1980) were primarily a result of infection. The development of acute leukemia has been reported in a small number of cases either early in the course of the disease or in adult long-term survivors (D'Oelsnitz et al., 1975; Kreshvan et al., 1978; Wasser et al., 1978; Alter, 1980; Basso et al., 1981). In general, leukemic transformation seems to be a rather rare event in the course of congenital hypoplastic anemia. With the advances in blood transfusion methods and in iron-chelation therapy, and most important with the improving experience in the use of corticosteroids, the majority of children with CHA are expected to reach adulthood.

Epilogue

It has been more than 60 years since the first description of pure red cell aplasia appeared in the medical literature. During these years this syndrome, despite its rarity, has attracted the interest of a great number of investigators whose studies have allowed a better definition of PRCA, the separation of the acquired from the congenital form, an appreciation of its variable clinical course, and its association with other diseases. Most important, the pathogenesis of this disorder has been widely investigated, its autoimmune nature has been better defined, and the association of human parvovirus infections with the acute erythroblastopenia in chronic hemolytic anemias has been established. The application of this knowledge to the management of this disorder resulted in the use of a number of immunosuppressive treatment modalities that either alone or in combination are capable of restoring erythropoiesis to normal in the majority of patients. In addition this disorder has served as a model for the study of other hematologic diseases of suspected autoimmune pathogenesis, such as pure white cell or megakaryocytic aplasia, drug-induced marrow injury, and aplastic anemia associated with eosinophilic fasciitis. Despite all this progress there still are a number of questions regarding the pathogenesis, the management, and the prognosis of this syndrome that remain unanswered.

Antibody-mediated PRCA has been well demonstrated in a number of cases. However, the antigen that is recognized by the serum IgG inhibitor on erythroid cells remains unknown. Furthermore, the variable requirement for complement for the action of this IgG in vitro indicates that there may be more than one antigenic determinant on the membrane of erythroid cells that is attacked by this IgG. The question of the nature of this antigen is of theoretical as well as practical significance since it seems that whatever this antigen may be, it is very specific for erythroid cells and is intimately associated with, and possibly important for, the erythroid cell differentiation. Identification of this antigen(s) may provide significant information relevant to normal erythroid cell differentiation.

Although the role of T-lymphocytes in suppressing erythropoiesis in the PRCA of chronic lymphocytic leukemia of B or T cell type has been well established, their role in primary acquired PRCA remains unclear. The significance of a variety of T cell abnormalities detected in the blood of these patients is not yet fully understood and whether these represent an epiphenomenon or are related to the basic autoimmune process and the pathogenesis of this disease remains unknown.

The association of PRCA with thymomas is beyond any doubt. However, how thymomas induce immune dysregulation that preferentially leads to PRCA, myasthenia gravis, or hypogammaglobulinemia remains an enigma.

The occurrence of transient erythroblastopenia of childhood after a viral illness is a common observation among pediatricians, and the appearance of new cases in an epidemic fashion is challenging. The specific agent responsible for the viral illness preceding this syndrome has not yet been studied, but it is tempting to hypothesize that an infectious agent exists that can cause a viral illness and a normal immune response, which in selected individuals includes the erythroid cells in the marrow. Under these circumstances one wonders whether the factor that determines the development of erythroblastopenia is the nature of the infectious agent or the genetically determined phenotype of the erythroid cells of the affected children.

The demonstration that the human parvovirus (B19) is the cause of PRCA in children with congenital hemolytic anemias has opened a new area of investigation and generated a number of questions. The mechanism through which this virus attacks only the erythroid progenitors and leaves the other hematopoietic progenitors almost intact remains to be determined. In addition, it is very likely that this parvovirus may not be the only one capable of attacking human marrow cells resulting in a specific cytopenia or possibly pancytopenia. The role of viruses in inducing bone marrow failure syndromes has just begun to be investigated.

Congenital hypoplastic anemia has been well separated from acquired PRCA for many years. Yet its pathogenesis remains unknown. It has not yet been defined whether the very low erythroid progenitor cell growth is due to low numbers of progenitors, to low sensitivity to erythropoietin, or to the presence of suppressor cells. The recent demonstration that cyclosporin can induce remission of this anemia will most likely reactivate the interest of the students of this syndrome in its possible autoimmune nature.

For the clinician there is not yet any single parameter or test that can be used reliably in an individual patient to determine the preleukemic or not nature of the illness and the selection of predictably successful treatment. The decision to stop treatment after induction of remission is difficult and predicting which patient will or will not relapse is impossible.

Despite all the above unsolved problems, recent advances in the area of erythropoiesis, cellular immunology, and in therapeutics make the look into the future very promising. The erythropoietin gene has been cloned and expressed and pure recombinant erythropoietin is now available for a variety of studies not previously possible. Erythroid progenitors have been isolated in high purity and numbers that will allow studies on the biochemical events during erythropoietin-induced differentiation. Specific erythropoietin receptors on erythroid cells have now been detected and studying their density and properties in disease states will soon become possible. On the other hand the field of cellular immunology is expanding fast and interactions between different subsets of T cells, T and B cells, or cytotoxic T cells and target cells can now be described with more detail and at a molecular level. In the field of therapeutics the availability of cyclosporin has provided new means of interfering with autoimmune phenomena, and promising results have come out of the first clinical trials. It is, therefore, reasonable to expect that in the years to follow the pathogenesis of the heterogenous syndrome of "pure" red cell aplasia will be better understood and its treatment will be more predictably successful.

References

Abeloff MD, Waterbury L. 1974. Pure red cell aplasia and chronic lymphocytic leukemia. *Arch Intern Med* 134:721.

Abkowitz JL, Kadin ME, Powell JS, Adamson JW. 1986*a*. Pure red cell aplasia: lymphocyte inhibition of erythropoiesis. *Br J Haematol* 63:59.

Abkowitz JL, Powell JS, Nakamura JM, Kadin ME, Adamson JW. 1986*b*. Pure red cell aplasia: response to therapy with anti-thymocyte globulin. *Am J Hematol* 23:363.

Aggio MC, Zunini C. 1977. Reversible pure red cell aplasia in pregnancy. *N Engl J Med* 297:221.

Albahary C. 1977. Thymome, myasthénie curable, anémie hémolytique recidivante puis érythroblastopénie aiguë cortico-curable. *Nouv Presse Med* 6:659.

Albahary C, Homberg TC, Guillaumine T, Martin S, Boulangiez TP. 1972. Thymome, myasthénie et anémie hémolytique auto-immune. *Nouv Presse Med* 1:1931.

Alter BP. 1980. Childhood red cell aplasia. *Am J Pediatr Hematol Oncol* 2:121.

Alter BP, Nathan DG. 1979. Red cell aplasia in children. *Arch Dis Child* 54:263.

Altman AC, Gross S. 1983. Severe congenital hypoplastic anemia: transmission from a healthy female to opposite sex step siblings. *Am J Pediatr Hematol Oncol* 5:99.

Altman KI, Miller G. 1953. A disturbance of tryptophan metabolism in congenital hypoplastic anemia. *Nature* 172:868.

Andersen SB, Ladefoged J. 1963. Pure red cell anemia and thymoma. *Acta Haematol* 30:319.

Anderson MJ. 1982. The emerging story of a human parvovirus-like agent. *J Hyg* 89:1.

Anderson MJ, Davis LR, Hodgson J, Jones SE, Murtaza L, Pattison JR, Stroud CE, White JM. 1982. Occurrence of infection with a parvovirus-like agent in children with sickle cell anemia during a two-year period. *J Clin Pathol* 35:744.

Anderson MG, Higgins PG, Davis LR, Willman JS, Jones SE, Kidd IM, Pattison JR, Tyrrell DAJ. 1985. Experimental parvoviral infection in humans. *J Infect Dis* 152:257.

Arrowsmith WR, Burris MB, Segaloff A. 1953. Production of megaloblastic marrow by administration of cortisone in aplastic anemia with subsequent response to vitamin B_{12}: a relationship not previously described. *J Lab Clin Med* 42:778.

August CS, King E, Githens JH, McIntosh K, Humbert JR, Greensheer A, Johnson FB. 1976. Establishment of erythropoiesis following bone marrow transplantation in a patient with congenital hypoplastic anemia (Diamond-Blackfan syndrome) *Blood* 48:491.

Baar H. 1927. Progressive postinfektiöse erythrophthise. *Folia Haematol* 35:111.

Bach JF, Dardenne M. 1972. Absence d'hormone thymique dans le serum de souris NZB et NZB X NZW et de malades atteints de lupus érythremateux disseminé. *J Urol Nephrol* 78:994.

Bacicalupo A, Podesta M, Mingari MC, Moretta L, Piaggio G, Van Lint MT, Durand A, Marmont AM. 1981. Generation of CFU-C/suppressor T cells in vitro: an experimental model for immune-mediated marrow failure. *Blood* 57:491.

Bakker PM. 1955. Enkele opmerkingen bij twee geswellen van de thymus. *Ned Tijdschr Geneeskd* 98:386.

Balaban EP, Buchanan GR, Graham M, Frenkel EP. 1985. Diamond-Blackfan syndrome in adult patients. *Am J Med* 78:533.

Banisadre M, Ash RC, Ascensao JL, Kay NE, Zanjani ED. 1981. Suppression of erythropoiesis by mitogen-activated T lymphocytes in vitro. In *Experimental Hematology Today*, ed. SJ Baum, DG Ledney, A. Khan Basel: Springer-Karger, 151.

Barbui T, Bassan R, Viero P, Minetti B, Comotti B, Buelli M. 1984. Pure white cell aplasia treated by high dose intravenous immunoglobulin. *Br J Haematol* 58:555.

Barnes RD. 1965. Thymic neoplasms associated with refractory anemia. *Guy Hosp Rep* 114:73.

———. 1966. Refractory anemia with thymoma. *Lancet* ii:1464.

Baron RL, Lee JK, Sagel SS, Levitt RG. 1982. Computed tomography of the abnormal thymus. *Radiology* 142:127.

Barosi G, Baraldi A, Cazzola M, Spriano P, Magrini U. 1983. Red cell aplasia in myelofibrosis with myeloid metaplasia: a distinct functional and clinical entity. *Cancer* 52:1290.

Barre C, Marteau J, Erlinger S, Berthaux P, Vignalou J. 1964. Tumeur thymique et anémie. *Presse Med* 72:2443.

Basso G, Cocito MG, Rebuffi L, Donzelli F, Milanesi C, Zanesco L. 1981. Congenital hypoplastic anaemia developed in acute megakaryoblastic leukaemia. *Helv Paediatr Acta* 36:267.

Batata MA, Martini N, Huvos AG. 1974. Thymomas: clinicopathologic features, therapy, and prognosis. *Cancer* 34:389.

Battle JD, Hewlett JS, Hoffman GC. 1963. Prolonged erythroid aplasia in chronic lymphocytic leukemia: favorable response to adrenocortical steroids in four cases. *Ann Intern Med* 58:731.

Baudouin J, Dezile G, Jobard P, Lavandier M, Jacquet J. 1968. Thymome et érythroblastopénie: analyse de 57 observations. *Presse Med* 76:13.

Beard MJE, Krantz SB, Johnson SAN, Bateman CJT, Whitehouse JMA. 1978. Pure red cell aplasia. *Q J Med* 187:339.

Bennett JM, Catovsky D, Daniel MT, Flandrin G, Galton DA, Gralnick HR, Sultan C. 1982. Proposals for the classification of the myelodysplastic syndromes. *Br J Haematol* 51:189.

Bentley SA, Murray KH, Lewis SM, Roberts PD. 1977. Erythroid hypoplasia in myelofibrosis: a feature associated with blastic transformation. *Br J Haematol* 36:41.

Bergh NP, Gatzinsky P, Larsson S, Lundin P, Ridell B. 1978. Tumors of the thymus and thymic region: I. clinicopathological studies on thymomas. *Ann Thorac Surg* 25:91.

Berlin G, Lieden G. 1986. Long-term remission of pure red cell aplasia after plasma exchange and lymphocytopheresis. *Scand J Haematol* 36:121.

Berlin NI, Lawrence JH, Lee HC. 1954. The pathogenesis of the anemia of chronic leukemia: measurement of life span of the red blood cells with glycine-C^{14}. *J Lab Clin Med* 44:860.

Berliner N, Duby AD, Linch DC, Murre C, Quertermous T, Knott LJ, Azin T, Newland AC, Lewis DL, Galvin MC, Seidman JG. 1986. T-cell receptor gene rearrangements define a monoclonal T cell proliferation in patients with T cell lymphocytosis and cytopenia. *Blood* 67:914.

Bernard J. 1969. Les aplasies pré-leucémiques. *Nouv Rev Fr Haematol* 9:41.

Bernard J, Najean Y, Slama R. 1962*a*. Tumeur de thymus avec érythroblastopénie: efficacité remarquable de la corticotherapie sur le foctionnement érythropoietique. *Nouv Rev Fr Haematol* 2:776.

Bernard J, Seligmann M, Chassigneux J, Dresch C. 1962*b*. Anémie de Blackfan-Diamond. *Nouv Rev Fr Haematol* 2:721.

Berner YN, Berrebi A, Green L, Handzel ZT, Bentwich Z. 1983. Erythroblastopenia in acquired immunodeficiency syndrome (AIDS). *Acta Haematol* 70:273.

Biemer JJ, Taylor FM. 1982. Transient reticulocytopenia in viral illness. *Ann Clin Lab Sci* 12:194.

Bjorkholm M, Holm G, Mellstedt H, Carberger G, Nisell J. 1976. Membrane bound IgG on erythroblasts in pure red cell aplasia following thymectomy: case report. *Scand J Haematol* 17:341.

Blanc M. 1980. Association leucémie lymphoide chroniques, cancer du rein et érythroblastopénie. *Nouv Presse Med* 9:882.

Bloom GE, Warner S, Gerald PS, Diamond LK. 1966. Chromosome abnormalities in constitutional aplastic anemia. *N Engl J Med* 274:8.

Boivin P, Eoche-Duval C. 1965. Une technique simple de dosage de l'érythropoietine chez la souris rendue polycythemique par hypoxie. *Rev Fr Etud Clin Biol* 10:434.

Bone marrow aplasia and parvovirus (editorial). 1983. *Lancet* ii:21.

Boston B. 1976. Chemotherapy of invasive thymoma. *Cancer* 38:49.

Bottcher D, Maas D, Wendt F, Schubothe H. 1970. Die Anämie durch Erythroblastopenie im Erwächsenenalter. *Klin Wochenschr* 48:96.

Böttiger LE, Rausing A. 1972. Pure red cell anemia: immunosuppressive treatment. *Ann Intern Med* 76:593.

Bourgeois-Droin C, Sauvanet A, Lemarchand F, Roquancourt A, Cottenot F, Brocherion C. 1981. Thymome, myasthénie, érythoblastopénie, myosite et myocardite a cellules géantes. *Nouv Presse Med* 10:2097.

Bouroncle BA. 1969. Familial crises in hereditary spherocytosis. *J Am Med Wom Assoc* 19:1045.

Bove JR. 1956. Combined erythroid hypoplasia and symptomatic hemolytic anemia. *N Engl J Med* 255:135.

Bowdler AJ. 1961. Radio-isotope investigations in primary myeloid metaplasia. *J Clin Pathol* 14:595.

Bowie PR, Teixeira OHP, Carpenter B. 1979. Malignant thymoma in a nine-year-old boy presenting with pleuropericardial effusion. *J Thorac Cardiovasc Surg* 77:777.

Brafield AJ, Verbov J. 1966. A case of thrombocythemia with red cell aplasia. *Postgrad Med J* 42:525.

Brittingham TE, Lutcher CL, Murphy DL. 1964. Reversible erythroid aplasia induced by diphenylhydantoin. *Arch Intern Med* 113:764.

Broccia G, Dessalvi P. 1983. Acute pure red cell aplasia following bromosulphophthalein injection in a patient with non-Hodgkin lymphoma. *Haematologica* 68:680.

Brody J. 1982. Diamond-Blackfan syndrome in father and son. *Aust Paediatr J* 18:128.

Brookfield EG, Singh P. 1974. Congenital hypoplastic anemia associated with hypogammaglobulinemia. *J Pediatr* 85:529.

Browman GP, Freedman MH, Blajchman MA, McBride JA. 1976. A complement independent erythropoietic inhibitor acting on the progenitor cell in refractory anemia. *Am J Med* 61:572.

Brunner HE. 1965. Die Osteomyelofibrose: Untersuchung der Ferro—und Erythrocyten—kinetik mit radioaktivem Eisen und Chrom. *Acta Haematol* 34:257.

Bruntsch U, Gallmeier WM, Burkhardt R, Schmidt CG. 1974. Immunosuppressive Therapie bei Erythroblastophthise. *Dtsch Med Wochenschr* 100:102.

Bruyn GA, Schelfhout LJ. 1986. Transient syndrome of inappropriate antidiuretic hormone secretion in a patient with infectious mononucleosis and pure red cell aplasia. *Neth J Med* 29:167.

Buchanan GR, Boxer LA, Nathan DG. 1976. The acute and transient nature of idiopathic immune hemolytic anemia in childhood. *J Pediatr* 88:780.

Budman DR, Steinberg AD. 1977. Hematologic aspects of systemic lupus erythematosus: current concepts. *Ann Intern Med* 86:220.

Bunn HF, McNeil BJ, Rosenthal DS, Krantz SB. 1976. Bone marrow imaging in pure red cell aplasia. *Arch Intern Med* 136:1169.

Burgert EO, Jr, Kennedy RL, Pease GL. 1954. Congenital hypoplastic anemia. *Pediatrics* 13:218.

Bussel JB, Hilgartner MW. 1984. The use and mechanism of action of intravenous immunoglobulin in the treatment of immune haematologic disease. *Br J Haematol* 56:1.

Bynoe AG, Scott CS, Ford P, Roberts BE. 1983. Decreased T helper cells in the myelodysplastic syndrome. *Br J Haematol* 54:97.

Callard RE, Smith CM, Worman C, Linch D, Cawley JC, Beverley PCL. 1981.

Unusual phenotype and function of an expanded subpopulation of T cells in patients with haematopoietic disorders. *Clin Exp Immunol* 43:497.

Carloss HW, Saab GA, Tavassoli M. 1979. Pure red cell aplasia and lymphoma. *JAMA* 242:67.

Cassileth PA, Myers AR. 1973. Erythroid aplasia in systemic lupus erythematosus. *Am J Med* 55:706.

Cathie IAB. 1950. Erythrogenesis imperfecta. *Arch Dis Child* 25:313.

Catovsky D, Lauria F, Matutes E, Foa R, Mantovin V, Tura S, Galton DA. 1981. Increase in T-gamma lymphocytes in B-cell chronic lymphocytic leukemia: II. correlation with clinical stage and findings in B prolymphocytic-leukemia *Br J Haematol* 47:539.

Cavalcant J, Shadduck RK, Winkelstein A, Zeigler Z, Mendelow H. 1978. Red cell hypoplasia and increased bone marrow reticulin in systemic lupus erythematosus: reversal with corticosteroid therapy. *Am J Hematol* 5:253.

Celada A, Farquet JJ, Muller AF. 1977. Refractory siberoblastic anemia secondary to autoimmune haemolytic anemia *Acta Haematol* 58:213.

Chalmers JNM, Boheimer K. 1954. Pure red cell anemia in patients with thymic tumors. *Br Med J* 2:1514.

Chan HS, Saunders EF, Freedman MH. 1982. Diamond Blackfan syndrome: I. erythropoiesis in prednisone-responsive and -resistant disease. *Pediatr Res* 16:474.

Chanarin I. 1976. Investigation and management of megaloblastic anemia. *Clin Haematol* 5:747.

Chanarin I, Burman D, Bennett MC. 1962. The familial aplastic crisis in hereditary spherocytosis: urocanic acid and formiminoglutamic acid excretion studies in a case with megaloblastic arrest. *Blood* 20:33.

Chang JC, Slutzker B, Lindsay N. 1978. Remission of pure red cell aplasia following oxymetholone therapy. *Am J Med Sci* 275:345.

Chapin MA. 1965. Benign thymoma, refractory anemia and hypogammaglobulinemia: case report. *J Maine Med Assoc* 56:83.

Charney E, Miller G. 1964. Reticulocytopenia in sickle cell disease. *Am J Dis Child* 107:450.

Chediak AB, Fuste R, Rosales GV. 1953. Timoma y anemia aplastica: consideraciones generales—presentacion de un caso. *Arch Hosp Univ* (Havana) 5:27.

Chernoff AI, Josephson AM. 1951. Acute erythroblastopenia in sickle cell anemia and infectious mononucleosis. *Am J Dis Child* 82:310.

Chikkappa G, Pasquale D, Mangan K, Tsan MF. 1986*a*. Successful treatment of pure red cell aplasia in a patient with chronic lymphocytic leukemia with cyclosporin. *Blood* 68 (suppl 1):106a.

Chikkappa G, Zarrali MH, Tsan MF. 1986*b*. Pure red cell aplasia in patients with chronic lymphocytic leukemia. *Medicine* 65:339.

Chollet P, Cavaroc M, de Laguillaumie J, Yerma TC, Rey M. 1973. Erythroblastopénie et lupus érythemateux disseminé. *Nouv Presse Med* 3:32.

Chollet P, Plagne R, Fonck Y, Chassagne J, Betoul G, Dardenne M, Bach JF. 1981. Thymoma with hypersecretion of thymic hormone. *Thymus* 3:321.

Claiborne RA, Dutt AK. 1985. Isoniazid-induced pure red cell aplasia. *Am Rev Respir Dis* 131:947.

Clark DA, Dessypris EN, Krantz SB. 1984. Studies on pure red cell aplasia: XI. results of immunosuppressive treatment of 37 patients. *Blood* 63:227.

Clarkson BC, Prockop DJ. 1958. Aregenerative anemia associated with benign thymoma. *New Engl J Med* 259:253.

Clauvel JP, Vainchenker W, Herrera A, Dellagi K, Vinci G, Tabilio A, Lacombe C. 1983. Treatment of pure red cell aplasia by high-dose intravenous immunoglobulins. *Br J Haematol* 55:380.

Clement F. 1979. Erythroblastopénie chronique acquise suivie de myélofibrose: rémissions sous traitement immunosuppresseur. *Nouv Presse Med* 8:3817.

Cohen DJ, Loertscher R, Rubin MF, Tilney NL, Carpenter CB, Strom TB. 1984. Cyclosporine: a new immunosuppressive agent for organ transplantation. *Ann Intern Med* 101:667.

Conley CL, Lippman SM, Ness PM. 1980. Autoimmune hemolytic anemia with reticulocytopenia: a medical emergency. *JAMA* 244:1688.

Conley CL, Lippman SM, Ness PM, Petz L, Branch DR, Gallagher MT. 1982. Autoimmune hemolytic anemia with reticulocytopenia and erythroid marrow. *N Engl J Med* 306:281.

Cornaglia-Ferraris P, Ghio R, Mori P, de Bernardi B, Pasino M, Sitia R, Massimo L. 1981. T-gamma lymphocytes in a case of congenital hypoplastic anemia *Haematologica* 66:289.

Cornet A, Cornu P, Barbier JP, Favriel JM, Massouline G. 1974. Un cas d'érythroblastopénie réversible due au thiamphenicol. *Sem Hop Paris* 50:1567.

Corrigan TT. 1981. Transient erythroblastopenia of childhood. *Ariz Med* 38:436.

Cossart YE, Field AM, Cant B, Widdow SD. 1975. Parvovirus-like particles in human sera. *Lancet* 1:72.

Couespel R, Gaillard JA, Vaillant G. 1962. Tumeur de thymus, anémie et hypoplasie érythroblastique. *Presse Med* 70:926.

Cowan RJ, Maynard CD, Witcofski RL, Janeway R, Toole JF. 1971. Selenomethionine S^{75} thymus scans in myasthenia gravis. *JAMA* 215:978.

Croles JJ, van Elden L. 1975. Immunological and electron microscopic observations in a case of pure red cell aplasia. *Neth J Med* 18:279.

Crosby WH. 1953. Paroxysmal nocturnal hemoglobinuria: report of a case complicated by an aregenerative (aplastic) crisis. *Ann Intern Med* 39:1107.

Crosby WH, Rappaport H. 1956. Reticulocytopenia in autoimmune hemolytic anemia. *Blood* 11:929.

Dainiak N, Hardin J, Floyd V, Callahan M, Hoffman R. 1980. Humoral suppression of erythropoiesis in systemic lupus erythematosus and rheumatoid arthritis. *Am J Med* 69:537.

Dalakas MC, Engel WK, McClure JE, Goldstein AL, Askanas V. 1981. Immunocytochemical localization of thymosin-a_1, in thymic epithelial cells of normal and myasthenia gravis patients and in thymic cultures. *J Neurol Sci* 50:239.

Dameshek W. 1941. Familial hemolytic crisis: report of three cases occurring within 10 days. *N Engl J Med* 224:52.

Dameshek W, Bloom ML. 1948. Events in hemolytic crisis of hereditary spherocytosis with particular reference to reticulocytopenia, pancytopenia and abnormal splenic mechanism. *Blood* 3:1381.

Dameshek W, Brown SM, Rubin AD. 1967. Pure red cell anemia (erythroblastic hypoplasia) and thymoma. *Semin Hematol* 4:222.

Daugaard G, Hansen HH, Rorth M. 1983. Combination chemotherapy for malignant thymoma. *Ann Intern Med* 99:189.

Daughaday WH. 1968. Clinicopathologic conference: lupus erythematosus with severe anemia, selective erythroid hypoplasia and multiple red blood cell isoantibodies. *Am J Med* 44:590.

Dausset J, Colombani J. 1959. The serology and prognosis of 128 cases of autoimmune hemolytic anemia. *Blood* 14:1280.

Davidshon I. 1941. Hemochromatosis, thymoma, severe anemia and endocarditis in a woman. Clinicopathologic conference. *Ill Med J* 80:427.

Davis LJ, Kennedy AC, Baikie AC, Brown A. 1952. Hemolytic anemias of various types treated with ACTH and cortisone: report of ten cases including one of acquired type in which erythropoietic arrest occurred during a crisis. *Glasgow Med J* 33:263.

deAlarcon PA, Miller ML, Stuart MJ. 1978. Erythroid hypoplasia: an unusual presentation of childhood acute lymphoblastic leukemia. *Am J Dis Child* 132:763.

Debusscher L, Paridaeus R, Stryckmans P, Delwiche F. 1985. Cyclosporine for pure red cell aplasia. *Blood* 65:249.

DeClerck YA, Ettenger RB, Ortega JA, Pennisi AJ. 1980. Macrocytosis and pure RBC anemia caused by azathioprine. *Am J Dis Child* 134:377.

Delfraissy JF, Tchernia G, Laurian Y. 1985. Suppressor cell function after intravenous gamma globulin treatment in adult chronic idiopathic thrombocytopenic purpura. *Br J Haematol* 60:315.

DeSevilla E, Forrest JV, Zirmuska FR, Sagel SS. 1975. Metastatic thymoma with myasthenia gravis and pure red cell aplasia. *Cancer* 36:1154.

Dessypris EN, Baer MR, Sergent JS, Krantz SB. 1984. Rheumatoid arthritis and pure red cell aplasia. *Ann Intern Med* 100:202.

Dessypris EN, Fogo A, Russell M, Engel E, and Krantz SB. 1980. Studies on pure red cell aplasia: X. association with acute leukemia and significance of bone marrow karyotype abnormalities. *Blood* 56:421.

Dessypris EN, Krantz SB. 1984. Effect of pure erythropoietin on DNA-synthesis by human marrow day 15 erythroid burst forming units in short term liquid culture. *Br J Haematol* 56:295.

———. 1985*a*. Aplastic anemia. In *Conn's Current Therapy*, ed. RE Rakel, Philadelphia: Saunders, 243.

———. 1985*b*. Primary refractory anemia: a clinical and laboratory study of erythropoiesis in 16 cases. *Am J Med Sci* 298:229.

Dessypris EN, Krantz SB, Roloff JS, Lukens JN. 1982. Mode of action of the IgG inhibitor of erythropoiesis in transient erythroblastopenia of childhood. *Blood* 59:114.

Dessypris EN, McKee LC, Metzantonakis C, Teliacos M, Krantz SB. 1981. Red cell aplasia and chronic granulocytic leukemia. *Br J Haematol* 48:217.

Dessypris EN, Redline S, Harris JW, Krantz SB. 1985. Diphenylhydantoin-induced pure red cell aplasia. *Blood* 65:789.

Diamond LK. 1978. Congenital hypoplastic anemia: Diamond-Blackfan syndrome: historical and clinical aspects. *Blood Cells* 4:209.

Diamond LK, Allen DM, Magill FB. 1961. Congenital erythroid hypoplastic anemia. *Am J Dis Child* 102:403.

Diamond LK, Blackfan KD. 1938. Hypoplastic anemia. *Am J Dis Child* 56:464.

Diamond LK, Wang WC, Alter BP. 1976. Congenital hypoplastic anemia. *Adv Pediatr* 22:349.

Di Benedetto T, Padre-Mendoza T, Albala MM. 1979. Pure red cell hypoplasia associated with long arm deletion of chromosome 5. *Hum Genet* 46:345.

DiGiacomo J, Furst SW, Nixon DD. 1966. Primary acquired red cell aplasia in the adult. *J Mount Sinai Hosp* (N.Y.) 33:382.

Di Gulielmo R, Miliani A, Citi S. 1963. Eritroblastoftisi aequisita: studio clinico, citomorfologico, umorale e radioisotopico. *Boll Soc Ital Ematol* 11:35.

D'Oelsnitz M, Vincent L, De Swarte M. 1975. A propos d'un cas de leucémie aiguë lymphoblastique survenue après guérison d'une maladie de Blackfan-Diamond. *Arch Fr Pediatr* 32:582.

Doll DC, Weiss RB. 1985. Neoplasia and the erythron. *J Clin Oncol* 3:429.

Donnelly MA. 1953. A case of primary red cell aplasia *Br Med J* i:438.

Douchain F, Leborgne JM, Robin M, Macart M, Kremp L. 1980. Erythroblastopénie aiguë par intolerance au dipropyl-acetate de sodium: premiere observation. *Nouv Presse Med* 9:1715.

Dreyfus B, Aubert P, Patte D, Fray A, Le Bolloc H, Combrisson A. 1962. Erythroblastopénie chronique avec tumeur de thymus: analyse de 43 observations. *Nouv Rev Fr Haematol* 2:739.

Dreyfus B, Aubert P, Patte D, Le Bolloc H, Combrisson A. 1963. Erythroblastopénie chronique découverte après une thymectomie: heureux effets de la corticothérapie a forte dose après èchee de doses moyennes. *Nouv Rev Fr Hematol* 3:765.

Duncan JR, Potter CB, Cappelini MD, Anderson MJ, Kurtz JB, Weatherall DJ. 1983. Aplastic crisis due to parvovirus infection in pyruvate kinase deficiency. *Lancet* ii:14.

Dunn AM, Kerr GD. 1981. Pure red cell aplasia associated with sulphasalazine. *Lancet* ii:1288.

Dwyer JM. 1987. Intravenous therapy with gamma globulin. *Adv Intern Med* 32:111.

Eaves CJ, Eaves ACE. 1985. Erythropoiesis. In *Hematopoietic Stem Cells*, ed. DW Golde, F Takaku. New York: Marcel Dekker, 19.

Eisemann G, Dameshek W. 1954. Splenectomy for pure red cell hypoplastic aregenerative anemia associated with autoimmune hemolytic disease: report of a case. *N Engl J Med* 251:1044.

Engstead L, Strauberg O. 1966. Hematological data and clinical activity of the rheumatoid disease. *Acta Med Scand* 180:13.

Entwistle CC, Fentem PH, Jacobs A. 1964. Red cell aplasia with carcinoma of the bronchus. *Br Med J* 2:1504.

Ershler WB, Ross J, Finlay JL, Shahidi NT. 1980. Bone marrow microenvironment defect in congenital hypoplastic anemia. *N Engl J Med* 302:1321.

Estavoyer JM, Singer P, Broda C, Baufle GH. 1981. Toxicité hématologique du thiamphenicol: a propos de deux observations d'érythroblastopénie aiguë. *Sem Hop Paris* 57:1970.

Etzioni A, Atias D, Pollack S, Levy J, Vardi P, Hazani H, Benderly A. 1986. Complete recovery of pure red cell aplasia by intramuscular gamma-globulin therapy in a child with hypoparathyroidism. *Am J Hematol* 22:409.

Evans WK, Thompson DM, Simpson WJ, Feld R, Phillips MJ. 1980. Combination chemotherapy in invasive thymoma: role of COPP. *Cancer* 46:1523.

Falter ML, Robinson MG. 1972. Autosomal dominant inheritance and aminoaciduria in Blackfan-Diamond anemia. *J Med Genet* 9:64.

Favre H, Chatelanat F, Miescher PA. 1979. Autoimmune hematologic diseases associated with infraclinical systemic lupus erythematosus in four patients: a human equivalent of the NZB mice. *Am J Med* 66:91.

Fehr J, Hofmann V, Kappeler U. 1982. Transient reversal of thrombocytopenia in idiopathic thrombocytopenic purpura by high-dose intravenous gamma globulin. *N Engl J Med* 306:1254.

Feldges D, Schmidt R. 1971. Erythrogenesis imperfecta (Blackfan-Diamond-Anaemie). *Schweiz Med Wochenschr* 101:1813.

Field EO, Caughi MN, Blackett NM, Smithers DW. 1968. Marrow suppressing factors in the blood in pure red cell aplasia, thymoma and Hodgkin's disease. *Br J Haematol* 15:101.

Fikrig SM, Berkovich S. 1969. Virus-induced aplastic crisis in mice. *Blood* 33:582.

Finch CA, Deubelbeiss K, Cook JD, Eschbach JW, Harker LA, Funk DD, Marsaglia G, Hillman RS, Slichter S, Adamson JW, Ganzoni A, Giblett ER. 1970. Ferrokinetics in man. *Medicine* 49:17.

Finkel HE, Kimber RJ, Dameshek W. 1967. Corticosteroid-responsive acquired pure red cell aplasia in adults. *Am J Med* 43:771.

Finlay JL, Shahidi NT, Horowitz S, Borcherding W, Hong R. 1982. Lymphocyte dysfunction in congenital hypoplastic anemia. *J Clin Invest* 70:619.

Fitzgerald PH, Hamer JW. 1971. Primary acquired red cell hypoplasia associated with a clonal chromosomal abnormality and disturbed erythroid proliferation. *Blood* 38:325.

Flury W, Montandon A. 1980. Isolierte erythroide Aplasie unter immunosuppressive Therapie mit Azathioprin nach Nierentransplantation. *Schweiz Med Wochenschr* 110:1614.

Fon GT, Bein ME, Mancuso AA, Keesey JC, Lupetin AR, Wong WS. 1982. Computed tomography of the anterior mediastinum in myasthenia gravis: a radiologic-pathologic correlative study. *Radiology* 142:135.

Fontain JR, Dales M. 1955. Pure red cell aplasia successfully treated with cobalt. *Lancet* i:541.

Förare SA. 1963. Pure red cell anemia in step siblings. *Acta Paediatr Scand* 52:159.

Fox RM, Firkin FC. 1978. Sequential pure red cell and megakaryocytic aplasia associated with chronic liver disease and ulcerative colitis. *Am J Hematol* 4:79.

Foy H, Kondi A. 1953*a*. A case of true red cell aplastic anemia successfully treated with riboflavin. *J Path Bact* 65:559.

———. 1953*b*. Primary red cell aplasia. *Br Med J* 1:1449.

Foy H, Kondi A, MacDougall L. 1961. Pure red cell aplasia in marasmus and kwashiorkor treated with riboflavin. *Br Med J* 1:937.

Foy H, Kondi A, Verjee ZH. 1972. Relation of riboflavin deficiency to corticosteroid metabolism and red cell hypoplasia in baboons. *J Nutr* 102:571.

Francis DA. 1982. Pure red cell aplasia: association with systemic lupus erythematosus and primary autoimmune hypothyroidism. *Br Med J* 284:85.

Freedman, MH. 1983. Recurrent erythroblastopenia of childhood: an IgM-mediated RBC aplasia. *Am J Dis Child* 137:458.

Freedman MH, Amato D, Saunders EF. 1975. Haem synthesis in the Diamond Blackfan syndrome. *Br J Haematol* 31:515.

———. 1976. Erythroid colony growth in congenital hypoplastic anemia. *J Clin Invest* 57:673.

Freedman MH, Saunders EF. 1978. Diamond-Blackfan syndrome: evidence against cell-mediated erythropoietic suppression. *Blood* 51:1125.

———. 1983. Transient erythroblastopenia of childhood: varied pathogenesis. *Am J Haematol* 14:247.

Freeman Z. 1960. Pure red cell anemia and thymoma: report of a case. *Br Med J* 1:1390.

Freund LG, Hippe E, Strandgaard S, Pelus LM, Erslev AJ. 1985. Complete remission in pure red cell aplasia after plasmapheresis. *Scand J Haematol* 35:315.

Frøland SS, Wisloff F, Stavem P. 1976. Abnormal lymphocyte populations in pure red cell aplasia. *Scand J Haematol* 17:241.

Funkhouser JW. 1961. Thymoma associated with myocarditis and the L.E. cell phenomenon. *N Engl J Med* 264:34.

Gajwani B, Zinner EN. 1976. Pure red cell aplasia associated with adenocarcinoma of stomach. *NY State J Med* 76:2177.

Gascon P, Zoumbos NC, Young N. 1984. Immunologic abnormalities in patients receiving multiple transfusions. *Ann Intern Med* 100:173.

Gasser C. 1949. Akute Erythroblastopenie: 10 Fälle aplastischer Erythroblastenkrisen mit Riesenproerythroblasten bei allergisch toxischen Zustandsbildern. *Helv Paediatr Acta* 4:107.

———. 1951. Aplastiche Anämie (chronische Erythroblastophthise) und Cortison. *Schweiz Med Wochenschr* 81:1241.

———. 1955. Pure red cell anemia due to auto-antibodies: immune-type of aplastic anemia, erthroblastopenia. *Sang* 26:6.

———. 1957. Aplasia of erythropoiesis. *Pediatr Clin North Am* 4:445.

Geary CG. 1978. Drug-related red cell aplasia. *Br Med J* i:51.

Geary CG, Byron PR, Taylor G, MacIver JE, Zervas J. 1975. Thymoma associated with pure red cell aplasia, immunoglobulin deficiency and an inhibitor of antigen-induced lymphocyte transformation. *Br J Haematol* 29:479.

Geller G, Drivit W, Zalusky R, Zanjani ED. 1975. Lack of erythropoietic inhibitory effect of serum from patients with congenital pure red cell aplasia. *N Engl J Med* 292:198.

Gerrits GP, van Oostrom CG, de Vaan GA, Bakkereu JA. 1984. Transient erythroblastopenia of childhood: a review of 22 cases. *Eur J Pediatr* 142:266.

Giannone L, Kugler JW, Krantz SB. 1987. Pure red cell aplasia associated with administration of sustained-release procainamide. *Arch Intern Med* 147:1179.

Gilbert EF, Harley JB, Anido V, Mengoli HF, Hughes JT. 1968. Thymoma, plasma cell myeloma, red cell aplasia and malabsorption syndrome. *Am J Med* 44:820.

Gill MJ, Ratliff DA, Harding LK. 1980. Hypoglycemic coma, jaundice and pure RBC aplasia following chlorpropamide therapy. *Arch Intern Med* 140:714.

Gill PJ, Amare M, Larsen WE. 1977. Pure red cell aplasia: three cases responding to immunosuppression. *Am J Med Sci* 273:213.

Girot R, Griscelli C. 1977. Erythroblastopénie a rechutes: a propos d'un cas suivi pendant 22 ans. *Nouv Rev Fr Haematol* 18:555.

Goldstein C, Pechet L. 1965. Chronic erythrocytic hypoplasia following pernicious anemia. *Blood* 25:31.

Gollan JL, Hussein S, Hoffbrand AV, Sherlock S. 1976. Red cell aplasia following prolonged d-penicillamine therapy. *J Clin Pathol* 29:135.

Goodman SB, Block MH. 1964. A case of red cell aplasia occurring as a result of antituberculous therapy. *Blood* 24:616.

Gordon RR, Varadi S. 1962. Congenital hypoplastic anemia (pure red cell aplasia) with periodic erythroblastopenia. *Lancet* i:296.

Gratwohl A, Speck B, Osterwalder M, Corneo M, Nissen C. 1983. Behandlung der Erythroaplasie (PRCA) mit antilymphozyten Globulin und kurzfristig hochdosierten Prednisone. *Schweiz Med Wochenschr* 113:1480.

Gray PH. 1982. Pure red cell aplasia: occurrence in three generations. *Med J Aust* 1:519.

Greco B, Beilory L, Stephanie D, Hsu SM, Gascon AN, Young N. 1983. Antithymocyte globulin reacts with many normal human cell types. *Blood* 62:1047.

Green P. 1958. Aplastic anemia associated with thymoma: report of two cases. *Can Med Assoc J* 78:419.

Greenberg PL, Nichols WC, Schrier SL. 1971. Granulopoiesis in acute myeloid leukemia and preleukemia. *N Engl J Med* 284:1225.

Guerra L, Najman A, Homberg JC, Duhamel G, Andre R. 1969. Auto-anticorps multiples au cours d'une maladie de Hodgkin: interprétation d'une érythroblastopénie. *Nouv Rev Fr Haematol* 9:601.

Guillan RA, Zelman S, Smalley RL, Iglesias PA. 1971. Malignant thymoma associated with myasthenia gravis and evidence of extrathoracic metastasis. *Cancer* 27:823.

Guthrie TH, Thornton RM. 1983. Pure red cell aplasia obscured by a diagnosis of carcinoma. *South Med J* 76:532.

Haas RJ, Netzel B, Meyer G, Huber W, Manz H. 1985. Anti-thymozyten-globulin Therapie bei schwerer aplastischer Anämie: eine Alternative zur Knochenmarks-transplantation. *Klin Padiatr* 197:101.

Hagberg H. Nilsson P, Nisell J. 1980. Treatment of pure red cell anemia with antilymphocytic globulin: report of two cases. *Scand J Haematol* 24:360.

Hamilton CR, Jr, Conley CL. 1969. Pure red cell aplasia and thymoma. *Johns Hopkins Med J* 125:262.

Hamilton PJ, Dawson AA, Galloway WH. 1974. Congenital erythroid hypoplastic anemia in mother and daughter. *Arch Dis Child* 49:71.

Hammond D, Keighley G. 1960. The erythrocyte-stimulating factor in serum and urine in congenital hypoplastic anemia. *Am J Dis Child* 100:466.

Hanada T, Abe T, Nakamura H, Aoki Y. 1984. Pure red cell aplasia: relationship between inhibitory activity of T cells to CFU-E and erythropoiesis. *Br J Haematol* 58:107.

Hanada T, Abe T, Takita H. 1985. T-cell mediated inhibition of erythropoiesis in transient erythroblastopenia of childhood. *Br J Haematol* 59:391.

Hankes LV, Brown RR, Schiffer L, Schmaler M. 1968. Tryptophan metabolism in humans with various types of anemias. *Blood* 32:649.

Hansen HG. 1951. Über die essentielle Erythroblastopenie. *Acta Haematol* 6:335.

Hansen RM, Lerner N, Abrams RA, Patrieck CW, Malik MI, Keller R. 1986. T-cell chronic lymphocytic leukemia with pure red cell aplasia: laboratory demonstration of persistent leukemia in spite of apparent complete remission. *Am J Hematol* 22:79.

Hanzawa A, Kuyama E, Hashiba T, Hayashi H. 1969. A case of pure red cell aplasia complicated with hypogammaglobulinemia. *Jpn J Clin Hematol* 10:630.

Harada M, Nakao S, Kondo K, Odaka K, Ueda M, Shiobara S, Matsue K, Mori T, Matsuda T. 1986. Effect of activated lymphocytes on the regulation of hematopoiesis: enhancement and suppression of in vitro BFU-E growth by T cells stimulated by autologous non-T cells. *Blood* 67:1143.

Hardisty RM. 1976. Diamond-Blackfan anemia. Ciba Foundation Symposium, no. 37 n.s. New York: Elsevier, 89.

Hare WSC, Andrews JT. 1970. The occult thymoma: radiological and radioisotopic aids to diagnosis. *Australas Ann Med* 19:30.

Harley J, Dods L. 1959. Some acquired hemolytic anemias of childhood. *Australas Ann Med* 8:98.

Harris SI, Weinberg JB. 1985. Treatment of red cell aplasia with antithymocyte globulin: repeated inductions of complete remissions in two patients. *Am J Hematol* 20:183.

Hartwell AS, Mermod LE. 1957. Erythroblastic hypoplasia associated with a thymoma: review and report of a case. *Hawaii Med J* 17:143.

Hauke G, Fauser AA, Weber S, Maas D. 1983. Reticutocytopenia in severe autoimmune hemolytic anemia (AIHA) of the warm antibody type. *Blut* 46:321.

Havard CW, Parrish JA. 1962. Thymic tumor and erythroblastic aplasia. *JAMA* 179:228.

Havard CW, Scott RB. 1960. Thymic tumor and erythroblastic aplasia: a report of 3 cases and a review of the syndrome. *Br J Haematol* 6:178.

Harvey AM, Schulman LE, Tumultz PA. 1954. Systemic lupus erythematosus: review of the literature and clinical analysis of 138 cases. *Medicine* 33:291.

Heck LW, Alarcon GS, Ball GV, Phillips RL, Kline LB, Moreno H, Hirschowitz BI, Baer AN, Dessypris EN. 1985. Pure red cell aplasia and protein-losing enteropathy in a patient with systemic lupus erythematosus. *Arthritis Rheum* 28:1059.

Hegde UM, Gordon-Smith EC, Worlledge SM. 1977. Reticulocytopenia and absence of red cell autoantibodies in immune hemolytic anemia. *Br Med J* 2:1444.

Heyn R, Kurczynski E, Schmickel R. 1974. The association of Blackfan-Diamond syndrome with physical abnormalities, and an abnormality of chromosome 1. *J Pediatr* 85:531.

Higasi T, Nakayama Y, Murata A, Nakamura K, Sugiyama M, Kawaguchi T, Suzuki S. 1972. Clinical evaluation of ^{67}Ga citrate scanning. *J Nucl Med* 13:196.

al-Hilali MA, Joyner MV. 1983. Pure red cell aplasia secondary to angioimmunoblastic lymphadenopathy. *J R Soc Med* 76:894.

Hinrichs VR, Stevenson TD. 1965. Thymoma, myasthenia gravis and aplastic anemia: report of a case. *Ohio State Med J* 61:31.

Hirai H. 1977. Two cases of erythroid hypoplasia caused by carbamazepine (Tegretol). *Rinsho Ketsueki* 18:33.

Hirst E, Robertson TI. 1967. The syndrome of thymoma and erythroblastopenic anemia: a review of 56 cases including 3 case reports. *Medicine* 46:225.

Hocking WG, Singh R, Schroff R, Golde DW. 1983. Cell mediated inhibition of erythropoiesis and megaloblastic anemia in T-cell chronic lymphocytic leukemia. *Cancer* 51:631.

Hoffman R, Kopel S, Hsu SD, Dainiak N, Zanjani ED. 1978. T cell chronic lymphocytic leukemia: presence in bone marrow and peripheral blood of cells that suppress erythropoiesis in vitro. *Blood* 52:255.

Hoffman R, McPhedram P, Benz EJ, Duffy TP. 1983. Isoniazid-induced pure red cell aplasia. *Am J Med Sci* 286:2.

Hoffman R, Young N, Ershler WB, Mazur E, Gewirtz A. 1982. Diffuse fasciitis and aplastic anemia: a report of four cases revealing an unusual association between rheumatologic and hematologic disorders. *Medicine* 61:373.

Hoffman R, Zanjani ED, Vila J, Zalusky R, Lutton J, Wasserman LR. 1976. Diamond-Blackfan syndrome: lymphocyte-mediated suppression of erythropoiesis. *Science* 193:899.

Holborow EJ, Asherson GL, Johnson GD, Barnes RD, Carmichael DS. 1963. Antinuclear factor and other antibodies in blood and liver diseases. *Br Med J* i:656.

Hotchkiss DJ, Jr. 1970. Pure red cell aplasia following pernicious anemia. *Proceedings of the 8th International Congress of Haematology*, Munich: JF Lehmans, 133.

Hotta T. Hirabayashi N, Kobayashi T, Nakamura S, Suzuki Y. 1980. Five cases showing hemoglobinuria and aplastic crisis following administration of diphenylhydantoin after craniotomy. *Rinsho Ketsueki* 21:536.

Houghton JB, Toghill PJ. 1978. Myasthenia gravis and red cell aplasia. *Br Med J* ii:1402.

Hughes DW. 1961. Hypoplastic anemia in infancy and childhood: erythroid hypoplasia. *Arch Dis Child* 36:349.

Huijgens PC, Thijs LG, den Ottolander GJ. 1978. Pure red cell aplasia, toxic dermatitis and lymphadenopathy in a patient taking diphenylhydantoin. *Acta Haematol* 59:31.

Humphreys GH, Southworth H. 1945. Aplastic anemia terminated by removal of a mediastinal tumor. *Am J Med Sci* 210:501.

Hunt FA, Lander CM. 1975. Successful use of combination chemotherapy in pure red cell aplasia associated with malignant lymphoma of histiocytic type. *Aust NZ J Med* 5:469.

Hunter C, Jacobs P, Richards J. 1981. Complete remission of idiopathic pure red cell aplasia. *S Afr Med J* 60:68.

Hunter RE, Hakami N. 1972. The occurrence of congenital hypoplastic anemia in half brothers. *J Pediatr* 81:346.

Hussain MAM, Flynn DM, Green N, Hussein S, Hoffbrand AV. 1976. Subcutaneous infusion and intramuscular injection of desferrioxamine in patients with transfusional iron overload. *Lancet* ii:1278.

Iannuci A, Perini A, Pizzolo G. 1983. Acquired pure red cell aplasia associated with thyroid carcinoma: a case report. *Acta Haematol* 69:62.

Ibrahim JM, Rawstron J, Booth J. 1966. A case of red cell aplasia in a Negro child. *Arch Dis Child* 41:213.

Imamura N, Kuramoto A, Morimoto T, Ihara A. 1986. Pure red cell aplasia associated with acute lymphoblastic leukemia of pre-T cell origin. *Med J Aust* 144:724.

Imamura S, Takigawa M, Ikai K, Yoshinaga H, Yamada M. 1978. Pemphigus foliaceus, myasthenia gravis, thymoma, and red cell aplasia. *Clin Exp Dermatol* 3:285.

Imbach P, D'Apuzzo V, Hirt A, Rossi E, Vest M, Barandrum S, Baumgartner C, Morell A, Shöni M, Wagner HP. 1981. High-dose intravenous gammaglobulin for idiopathic thrombocytopenic purpura. *Lancet* ii:475.

Inoue S, Ravindranath Y, Lusher JM, Ito T. 1980. Absence of inhibitory cells from patients with aplastic anemia or transient erythroblastopenia of childhood on the in vitro growth of erythroid colonies. *Acta Haematol Jpn (Nippon Ketsueki Gakkai Zasshi)* 43:76.

Iriondo A, Garijo J, Baro J, Conde E, Pastor JM, Sabanes A, Hermosa V, Sainz MC, Perez de la Lastra L, Zubizarreta A. 1984. Complete recovery of hemopoiesis following bone marrow transplant in a patient with unresponsive congenital hypoplastic anemia (Blackfan-Diamond syndrome). *Blood* 64:348.

Isbister JP, Ralston M, Hayes JM, Wright R. 1981. Fulminant lupus pneumonitis with acute renal failure and RBC aplasia. *Arch Intern Med* 141:1081.

Iskandar O, Jager MJ, Willenze R, Natarajan AT. 1980. A case of pure red cell aplasia with a high incidence of spontaneous chromosome breakage: a possible X-ray sensitive syndrome. *Hum Genet* 55:337.

Jacobs AD, Champlin RE, Golde DW. 1985. Pure red cell aplasia characterized by erythropoietic maturation arrest: response to anti-thymocyte globulin. *Am J Med* 78:515.

Jacobs EM, Hutter RV, Pool JL, Ley AB. 1959. Benign thymoma and selective erythroid aplasia of the bone marrow. *Cancer* 12:47.

Jahsman DP, Monto RW, Rebuck JW. 1962. Erythroid hypoplastic anemia (erythroblastopenia) associated with benign thymoma. *Am J Clin Pathol* 38:152.

Jeong YG, Yung Y, River GL. 1974. Pure red blood cell aplasia and diphenylhydantoin. *JAMA* 229:314.

Jepson JH, Lowenstein L. 1966. Inhibition of erythropoiesis by a factor present in the plasma of patients with erythroblastopenia. *Blood* 27:425.

Jepson JH, Vas M. 1974. Decreased in vivo and in vitro erythropoiesis induced by plasma of ten patients with thymoma, lymphosarcoma or idiopathic erythroblastopenia. *Cancer Res* 34:1325.

Josephs HW. 1936. Anemia in infancy and early childhood. *Medicine* 15:307.

Jurgensen JC, Abraham JP, Hardy WW. 1970. Erythroid aplasia after halothane hepatitis: report of a case. *Am J Dig Dis* 15:577.

Kaplan J, Sarnaik S, Gitlin J, Lusher J. 1984. Diminished helper/suppressor lymphocyte ratios and natural killer activity in recipients of repeated blood transfusions. *Blood* 64:308.

Kark RM. 1937. Two cases of aplastic anemia: one with secondary hemochromatosis following 290 transfusions in nine years, the other with secondary carcinoma of stomach. *Guy Hosp Rep* 87:343.

Katakkar SB. 1986. Pure red cell aplasia: response to intravenous immunoglobulin, a blocking antibody. *Arch Intern Med* 146:2288.

Kato M, Shirai T, Umeda M, Kiga Y, Maki K, Kaneko H, Watanabe S, Hirahata T, Ishikawa I, Takatsuki Y, Masaki T, Yamauchi M, Tsukahara T. 1979. A case of pure red cell aplasia which progressed to acute myelomonocytic leukemia after 8 months. *Rinsho Ketsueki* 20:1451.

Katz JL, Hoffman R, Ritchey AK, Dainiak N. 1981. The proliferative capacity of pure red cell aplasia bone marrow cells. *Yale J Biol Med* 54:89.

Kaznelson P. 1922. Zur Entstehung der Blutplattchen. *Verh Dtsch Ges Inn Med* 34:557.

Kelleher JF, Luban NL, Mortimer PP, Kamimura T. 1983. Human serum "parvovirus": a specific cause of aplastic crisis in children with hereditary spherocytosis. *J Pediatr* 102:720.

Kesse-Elias M, Gyftaki E, Alevizou-Terzaki V, Malamos B. 1968. ^{59}Fe and ^{51}Cr studies in aplastic anemia and myelosclerosis. *Acta Haematol* 39:139.

Khatua SP. 1964. Congenital hypoplastic anemia (pure red cell type). *Indian J Pediatr* 1:276.

Khelif A, Van VH, Tremisi JP, Alfonsi F, Perrot D, Motin J, Viala JJ. 1985. Remission of acquired pure red cell aplasia following plasma exchanges. *Scand J Haematol* 34:13.

Kho LK. 1957. Erythroblastopenia with giant proerythroblasts in kwashiorkor. *Blood* 12:171.

Kho LK, Odang O, Thajeb S, Markum AH. 1962. Erythroblastopenia (pure red cell aplasia) in childhood in Djakarta. *Blood* 19:168.

Kimura N, Yamaraki K, Niho Y, Kisu T, Oka Y, Otuka T. 1983. Pure red cell aplasia associated with juvenile rheumatoid arthritis: effect of T-lymphocytes on hemopoietic progenitor cells. *Acta Haematol Jpn* 45:129.

Kitahara M. 1978. Pure RBC aplasia associated with chronic granulocytic leukemia. *JAMA* 240:376.

Kloster J. 1934. Über atypische Anämien. *Folia Haematol* 51:251.

Koeffler HP, Golde DW. 1980. Human preleukemia. *Ann Intern Med* 93:347.

Koenig HM, Lightsey AL, Nelson DP, Diamond LK. 1979. Immune suppression of erythropoiesis in transient erythroblastopenia of childhood. *Blood* 54:742.

Kondi A, Foy H. 1964. Vacuolisation of early erythroblasts in riboflavine-deficient baboons and in marasmus and kwashiorkor. *Lancet* ii:1157.

Kondi A, MacDougall L, Foy H, Mehta SH, Mbaya V. 1963. Anemias of marasmus and kwashiorkor in Kenya. *Arch Dis Child* 38:267.

Konwalinka G, Huber C, Tomaschek B, Peschle C, Geissler D, Odavik R, Braunsteiner H. 1983. A case of acquired pure red cell anemia studied by cloning of erythroid progenitor cells in vitro. *Acta Haematol* 70:316.

Kough RH, Barnes WT, 1964. Thymoma associated with erythroid aplasia, bullous skin eruption and the lupus erythematosus cell phenomenon: report of a case. *Ann Intern Med* 61:308.

Kozuru M, Nakashima K, Noda Y, Umemura T, Ibayashi H. 1976. Pure red cell aplasia associated with hypogammaglobulinemia: unresponsiveness of the bone marrow cells to erythropoietin in vitro. *J Kyshu Hematol* 25:1.

Krantz SB. 1965. The effect of erythropoietin on human bone marrow cells in vitro. *Life Sci* 4:2393.

———. 1972*a*. Studies on red cell aplasia. III: treatment with horse antihuman thymocyte gamma globulin. *Blood* 39:347

———. 1972*b*. Studies on red cell aplasia. IV: treatment with immunosuppressive drugs. In *Regulation of Erythropoiesis*, ed. AS Gordon, M Condorelli, C Peschle. Rome: Il Ponte, 312.

———. 1974. Pure red cell aplasia. *N Engl J Med* 291:345.

———. 1976. Diagnosis and treatment of pure red cell aplasia. *Med Clin North Am* 60:945.

———. 1978. Implications of studies on pure red cell aplasia for the study of aplastic anemia. In *Aplastic Anemia*, ed. S Hibino. Baltimore: University Park Press, 305.

———. 1983. Pure red cell aplasia. In *Current Therapy in Hematology-Oncology, 1983–1984*, ed. MC Brain, PB McCulloch. Toronto: CV Mosby, 6.

———. 1987. Pure red cell aplasia. In *Surgery of the Thymus*, ed. JC Givel. Berlin-Heidelberg: SpringerVerlag, in press.

Krantz SB, Dessypris EN. 1981. The role of antibody in pure red cell aplasia. In *Proceedings of the Conference on Aplastic Anemia: A Stem Cell Disease*, ed. AS Levine. Washington, D.C.: NIH (NIH Publication 81-1008), 157.

———. 1982. Antibody-dependent cellular cytotoxicity to allogeneic but not autologous erythroblasts in vitro. *J Clin Immunol* 2:222.

———. 1985. Pure red cell aplasia. In *Hematopoietic Stem Cells*, ed. DW Golde, F Takaku. New York: Marcel Dekker, 229.

Krantz SB, Gallien-Lartique O, Goldwasser E. 1963. The effect of erythropoietin upon heme synthesis by marrow cells in vitro. *J Biol Chem* 238:4085.

Krantz SB, Goldwasser E. 1987. Antibody-mediated inhibition of erythropoiesis. In *Humoral and Cellular Regulation of Erythropoiesis*, ed. ED Zanjani, M Tavassoli, JL Ascensao. New York: P.M.A. Publishing Corp., in press.

Krantz SB, Jacobson LO. 1970. *Erythropoietin and the Regulation of Erythropoiesis*. Chicago: University of Chicago Press.

Krantz SB, Kao V. 1967. Studies on red cell aplasia. I: demonstration of a plasma inhibitor to heme synthesis and an antibody to erythroblast nuclei. *Proc Natl Acad Sci USA* 58:493.

———. 1969. Studies on red cell aplasia. II: report of a second patient with an antibody to erythroblast nuclei and a remission after immunosuppressive therapy. *Blood* 34:1.

Krantz SB, Moore WH, Zaentz SD. 1973. Studies on red cell aplasia. V: presence of erythroblast cytotoxicity in γG-globulin fraction of plasma. *J Clin Invest* 52:324.

Krantz SB, Zaentz SD. 1977. Pure red cell aplasia. In *The Year in Hematology—1977*, ed. AS Gordon, R Silber, J LoBue. New York: Plenum, 153.

Kreshvan EU, Wegner K, Gara SK. 1978. Congenital hypoplastic anemia terminating in acute promyelocytic leukemia. *Pediatrics* 61:898.

Krivoy N, Ben-Arieh Y, Carter A, Alroy G. 1981. Methazolamide-induced hepatitis and pure RBC aplasia. *Arch Intern Med* 141:1229.

Kubic PT, WarKentin PF, Levitt CJ. 1979. Transient erythroblastopenia of childhood occurring in clusters. *Pediatr Res* 13:435.

Kuriyama K, Tomonaga M, Jinnai I, Matsuo T, Yoshida Y, Amenomori T, Yamada Y, Ichimaru M. 1984. Reduced helper (OKT4+):suppressor (OKT8+) T ratios in aplastic anemia: relation to immunosuppressive therapy. *Br J Haematol* 57:329.

Kurrein F. 1959. Pure red cell anemia coincident with benign thymoma. *J Clin Path* 12:319.

Kurtzman GJ, Ozawa K, Cohen B, Hanson G, Oseas R, Young NS. 1987. Chronic bone marrow failure due to persistent B19 parvovirus infection. *N Engl J Med* 317:287.

Labotka RJ, Maurer HS, Honig GR. 1981. Transient erythroblastopenia of childhood: review of 17 cases including a pair of identical twins. *Am J Dis Child* 135:937.

Lacombe C, Casadevall N, Muller O, Varet B. 1984. Erythroid progenitors in adult chronic pure red cell aplasia: relationship of in vitro erythroid colonies to therapeutic response. *Blood* 64:71.

Lambie AT, Burrows, BA, Sommers SC. 1957. Clinicopathologic conference: refractory anemia, agammaglobulinemia and mediastinal tumor. *Amer J Clin Pathol* 27:444.

Lane M, Alfrey CP. 1965. The anemia of human riboflavin deficiency. *Blood* 25:432.

Lane M, Alfrey CP, Mengel CE, Doherty MA, Doherty J. 1964. The rapid induction of human riboflavin deficiency with galactoflavin. *J Clin Invest* 43:357.

Lattes R. 1962. Thymoma and other tumors of the thymus: an analysis of 107 cases. *Cancer* 15:1224.

Lauriola LL, Maggiano N, Marino M, Carbone A, Mauro P, Musiani P. 1981. Human thymoma: immunologic characteristics of the lymphocytic component. *Cancer* 48:1992.

Lawton JW, Aldrich JE, Turner TL. 1974. Congenital erythroid hypoplastic anemia: autosomal dominant transmission. *Scand J Haematol* 13:276.

Lebrun E, Ajchenbaum F, Troussard X, Galateau F, Leporrier M, Lacombe C, Casadevall N, Varet B, Vernant JP, Dumont J, Binet JL, Piette M, Dreyfus B. 1985. Leucémie lymphoide chronique, érythroblastopénie, thymolipome. *Nouv Rev Fr Haematol* 27:29.

Lee CH, Clark AR, Thorpe ME, Mackie BS, Firkin FC. 1980. Bile duct adenocarcinoma with Leser-Trelat sign and pure red blood cell aplasia. *Cancer* 46:1657.

Lee CH, Firkin FC, Grace CS, Rozenberg MC. 1978. Pure red cell aplasia: a report of three cases with studies on circulating toxic factors against erythroid precursors. *Aust NZ J Med* 8:75.

Legg MA, Brady WJ. 1965. Pathology and clinical behavior of thymomas: a survey of 51 cases. *Cancer* 18:1131.

LeGolvan DP, Abell MR. 1977. Thymomas. *Cancer* 39:2142.

Lehman G, Alcoff J. 1982. Reversible pure red cell hypoplasia in pregnancy. *JAMA* 247:1170.

Leiken SL. 1957. The aplastic crisis of sickle cell disease: occurrence in several members of families within a short period of time. *Am J Dis Child* 93:1289.

Lemenager J, Tanguy A, Bernard Y. 1973. Thymome avec déficit immunitaire, érythroblastopénie, paraproteine monoclonal et plasmacytose médullaire. *Nouv Presse Med* 2:1374.

Levine PH, Hamstra RD. 1969. Megaloblastic anemia of pregnancy simulating acute leukemia. *Ann Intern Med* 71:1141.

Levinson AI, Hoxie JA, Kornstein MJ, Zembryki D, Matthews DM, Schreiber AD. 1985. Absence of the OKT4 epitope on blood T cells and thymus cells in a patient with thymoma, hypogammaglobulinemia and red blood cell aplalsia. *J Allergy Clin Immunol* 76:433.

Levitt L, Moonka D, Engelman E, Cabradilla C. 1983*a*. HTLV-associated T-suppressor cell inhibition of erythropoiesis in a patient with pure red cell aplasia and hypogammaglobulinemia. *Blood* 62(suppl 1): 48a.

Levitt LJ, Ries CA, Greenberg PL. 1983*b*. Pure white cell aplasia: antibody-mediated autoimmune inhibition of granulopoiesis. *N Engl J Med* 308:1141.

Lewis, SM, Szur L. 1963. Malignant myeloscleosis. *Br Med J* ii:472.

Linch DC, Cawley JC, MacDonald SM, Masters G, Roberts BE, Antonis AH, Waters AK, Sieff C, Lydyard PM. 1981. Acquired pure red cell aplasia associated with an increase of T-cells bearing receptors for the Fc of IgG. *Acta Haematol* 65:270.

Link MP, Alter BP. 1981. Fetal-like erythropoiesis during recovery from transient erythroblastopenia of childhood. *Pediatr Res* 15:1036.

Linman JW, Bragby G, Jr. 1976. The preleukemic syndrome: clinical and laboratory features, natural course and management. *Blood Cells* 2:11.

Linsk JA, Murray CK. 1961. Erythrocyte aplasia and hypogammaglobulinemia: response to steroids in a young adult. *Ann Intern Med* 55:831.

Lippman SM, Durie BG, Garewal HS, Giordano G, Greenberg BR. 1986. Efficacy of danazol in pure red cell aplasia. *Am J Hematol* 23:373.

Lipton JM, Kudisch M, Gross R, Nathan DG. 1986. Defective erythroid progenitor differentiation system in congenital hypoplastic (Diamond-Blackfan) anemia. *Blood* 67:962.

Lipton JM, Nadler LM, Canellos GP, Kudisch M, Reiss CS, Nathan DG. 1983. Evidence for genetic restriction in the suppression of erythropoiesis by a unique subset of T-lymphocytes in man. *J Clin Invest* 72:694.

Litwin SD, Zanjani ED. 1977. Lymphocytes suppressing both immunoglobulin production and erythroid differentiation in hypogammaglobulinemia. *Nature* 266:57.

Loeb V. 1956. Presentation of a case: comment in diagnostic problems. *JAMA* 160:1319.

Loeb V, Moore CV, Dubach R. 1953. The physiologic evaluation and management of chronic bone marrow failure. *Am J Med* 15:499.

Lovric VA. 1970. Anaemia and temporary erythroblastopenia in children. *Aust Ann Med* I:34.

Lowenberg B, Ghio R. 1977. An assay for serum cytotoxicity against erythroid precursor cells in pure red cell aplasia. *Biomedicine* 27:285.

Lupu NC, Nicolau GCT. 1931. Observations cliniques, hématologiques et histologiques sur un cas de leucoérythrophthisie. *Sang* 5:530.

MacCulloch D, Jackson JM, Venerys J. 1974. Drug-induced red cell aplasia. *Br Med J* 4:163.

McCurdy PR. 1961. Chloramphenicol bone marrow toxicity. *JAMA* 176:588.

MacDougall-Lorna G. 1982. Pure red cell aplasia associated with sodium valproate therapy. *JAMA* 247:53.

McFarland G, Say B, Carpenter NJ, Plunket DC. 1982. A condition resembling congenital hypoplastic anemia occurring in a mother and son. *Clin Pediatr* 21:755.

McGrath BP, Ibels LS, Raik E, Hargrave M, Mahony JF, Stewart JH. 1975. Erythroid toxicity of azathioprine: macrocytosis and selective marrow hypoplasia. *Q J Med* 173:57

McGuire WA, Yang HH, Bruno E, Coates TD, Hoffman R. 1986. Successful treatment of antibody-mediated pure red cell aplasia with high-dose intravenous immunoglobulin therapy. *Blood* 68(suppl 1):113a.

MacIver JE, Parker-Williams EJ. 1961. The aplastic crises in sickle cell anemia. *Lancet* 1:1086.

MacKechnie HLN, Squires AH, Platts M, Pruzanski W. 1973. Thymoma, myasthenia gravis, erythroblastopenic anemia, and systemic lupus erythematosus in one patient. *Can Med Assoc J* 109:733.

McLoud TC, Wittenberg J, Ferucci JT. 1979. Computed tomography of the thorax and standard radiographic evaluation of the chest: a comparative study. *J Comput Assist Tomogr* 3:170.

MacMahon JN, Egan EL. 1980. Aplastic anemia in a patient with pure red cell aplasia. *Ir J Med Sci* 149:212.

Madder B, Foley JE, Fisher JW. 1978. The in vitro and in vivo effects of testosterone and steroid metabolites on erythroid colony forming cells (CFU-E). *J Pharmacol Exp Ther* 207:1004.

Mannoji M, Shimoda M, Koresawa S, Yamada O, Togawa A, Yawata Y, Umemura H, Kozuru M. 1981. A case of angioimmunoblastic lymphadenopathy with dysproteinemia associated with autoimmune hemolytic anemia and pure red cell aplasia with special references to its pathogenesis. *Rinsho Ketsueki* 22:1751.

Mangan KF, D'Alessandro L. 1985. Hypoplastic anemia in B-cell chronic lymphocytic leukemia: evolution of T-cell mediated suppression of erythropoiesis in early stage and late stage disease. *Blood* 66:533.

Mangan KF, Bera EC, Shaddner RK, Tedrow H, Ray PK. 1984. Demonstration of two distinct antibodies in autoimmune hemolytic anemia with reticulocytopenia and red cell aplasia. *Exp Hematol* 12:788.

Mangan KF, Chikkappa G, Farley PC. 1982. T-gamma (T_γ) cells suppress growth of erythroid colony forming units in vitro in the pure red cell aplasia of B-cell chronic lymphocytic leukemia. *J Clin Invest* 70:1148.

Mangan KF, Chikkappa G, Scharfman WB, Desforges JF. 1981. Evidence for reduced erythroid burst promoting function of T lymphocytes in the pure red cell aplasia of chronic lymphocytic leukemia. *Exp Hematol* 9:489.

Mangan KF, Volkin R, Winkelstein A. 1986. Autoreactive erythroid progenitor-T suppressor cells in the pure red cell aplasia associated with thymoma and pan-hypogammaglobulinemia. *Am J Hematol* 23:167.

Marinone G, Mombelloni P, Marini G, Ghio R, Roncoli B, Rossi G, Verdura P, Protto C. 1981. Bone marrow erythroblastic recovery after plasmapheresis in acquired pure red cell anemia: case report. *Hematologica* 66:796.

Marmont A. 1977. Pure red cell aplasia as an autoimmune receptor disease. *Blood* 49:155.

———. 1984. Is antilymphocytic globulin a better immunosuppressant than cyclophosphamide for pure red cell aplasia. *Br J Haematol* 56:680.

Marmont A, Gori E. 1973. Eritroblastopenia pura associata a timoma: remissione completa dopo terapia immunodepressiva. *Haematologica* 58:336.

Marmont A, Gerri R, Lereari G, Van Lint MT, Bacicalupo A, Risso M. 1985. Positive direct antiglobulin tests and heteroimmune hemolysis in patients with severe aplastic anemia and pure red cell anemia treated with antilymphocytic globulin. *Acta Haematol* 74:14.

Marmont A, Peschle C, Sanguinetti M, Condorelli M. 1975. Pure red cell aplasia: response of three patients to cyclophosphamide and/or antilymphocyte globulin (ALG) and demonstration of two types of serum IgG inhibitors to erythropoiesis. *Blood* 45:247.

Matras A, Priesel A. 1928. Uber einege Gewächse des Thymus. *Beitr path Anat u z allg Path* 80:270.

Meeus-Bith L, Boiron M, Paoletti C, Christol D, Tubiana M, Bousser J. 1957. Contribution de l'étude physiopathologique des splenomegalies myéloides avec myélofibrose. *Rev Hematol* 12:445.

Merrick MV, Gordon Smith EC, Lavender JP, Szur L. 1975. A comparison of ^{111}In with ^{59}Fe and ^{99}Tc-sulfur colloid for bone marrow scanning. *J Nucl Med* 16:66.

Messerschmitt J, Colonna P, Belkhodja A, Hamladji RM, Timsit G. 1971. Erythroblastopénie au cours d'une anémie de grossesse. *Nouv Rev Fr Haematol* 11:119.

Messner HA, Fauser AA, Curtis JE, Dotten D. 1981. Control of antibody-mediated pure red cell aplasia by plasmapheresis. *N Eng J Med* 304:1334.

Meyer LM, Bertcher RW. 1960. Acquired hemolytic anemia and transient erythroid hypoplasia of the bone marrow. *Am J Med* 28:606.

Meyer RJ, Hoffman R, Zanjani ED. 1978. Autoimmune hemolytic anemia and periodic pure red cell aplasia in systemic lupus erythematosus. *Am J Med* 65:342.

Michael SR, Vural IL, Bassen FA, Schaefer L. 1951. The hematologic aspects of disseminated (systemic) lupus erythematosus. *Blood* 6:1059.

Mielke HG. 1957. Aplastische Anämie (Erythroblastophthise) bei gutartigen Thymus Tumor. *Aerztl Wschr* 12:556.

———. 1958. Aplastische Anämie (Erythroblastophthise) nach INH-behandlung. *Folia Haematol* 2:1.

Migueres J, Paillas J, Ducos J, Jover A, Tremoulet M. 1968. Thymomes malignes et syndromes associés: myasthénie, carence en anticorps, syndromes d'autoimmunization, syndrome hématologique: a propos d'une observation. *Sem Hop Paris* 44:2809.

Miller DR. 1978. Erythropoiesis and hypoplastic anemias. In *Smith's Blood Diseases of Infancy and Childhood*, ed. DR Miller, HA Pearson, RL Baehner, CW McMillan. 4th ed. St. Louis: Mosby, 231.

Milner GR, Geary CG, Wadsworth LD, Doss A. 1973. Erythrokinetic studies as a guide to the value of splenectomy in primary myeloid metaplasia. *Br J Haematol* 25:467.

Milnes JP, Goorney BP, Wallington TB. 1984. Pure red cell aplasia and thymoma associated with high levels of the suppressor/cytotoxic T lymphocyte subset. *Br Med J* 289:1333.

Min KW, Waddell C, Pircher FJ, Granville GE, Gyokey F. 1978. Selective uptake of ^{75}Se-selenomethionine by thymoma with pure red cell aplasia. *Cancer* 41:1323.

Mitchell ABS, Pinn G, Pegrum GD. 1971. Pure red cell aplasia and carcinoma. *Blood* 37:594.

Miyoshi I, Hikita T, Koi B, Kimura I. 1978. Reversible pure red cell aplasia in pregnancy. *N Engl J Med* 299:777.

Mladenovic J, Farber N, Burton JD, Zanjani ED, Jacobs HS. 1985. Antibody to the erythropoietin receptor in pure red cell aplasia. *Blood* 66 (suppl 1):122a.

Moeschlin S, Rohr K. 1943. Aplastische Anämie mit jahrelangem vollständigem Fehlen der Erythroblasten (Erythroblastophthise). *Dtsch Arch Klin Med* 190:117.

Moloney WC. 1979. Case records of the Massachusetts General Hospital: case 40—1979. Scully RE, Gialdbini JJ, McNeely BU, Eds. *N Engl J Med* 301:770.

al-Mondhiry H, Zanjani ED, Spivack M, Zalusky R, Gordon AS. 1971. Pure red cell aplasia and thymoma: loss of serum inihibitor of erythropoiesis following thymectomy. *Blood* 38:576.

Morgan E, Pang KM, Goldwasser E. 1978. Hodgkin's disease and red cell aplasia. *Am J Hematol* 5:71.

Mortimer PP. 1983. Hypothesis: the aplastic crisis of hereditary spherocytosis is due to a single transmissible agent. *J Clin Pathol* 36:445.

Mortimer PP, Humphries KR, Moore JG, Purcell RH, Young NS. 1983*a*. A human parvovirus-like virus inhibits haematopoietic colony formation in vitro. *Nature* 302:426.

Mortimer PP, Luban NL, Kelleher JF, Cohen BJ. 1983*b*. Transmission of serum parvovirus-like virus by clotting factor concentrates. *Lancet* 2:482.

Mott MG, Apley J, Raper AB. 1969. Congenital (erythroid) hypoplastic anemia: modified expression in males. *Arch Dis Child* 44:757.

Murray WD, Webb JN. 1966. Thymoma associated with hypogammaglobulinemia and pure red cell aplasia. *Am J Med* 41:974.

Myers TJ, Bower FB, Hild DH. 1980. Pure red cell aplasia and the syndrome of multiple endocrine gland insufficiency. *Am J Med Sci* 280:29.

Nagasawa T, Abe T, Nakagawa T. 1981. Pure red cell aplasia and hypogammaglobulinemia associated with T_{γ}-cell chronic lymphocytic leukemia. *Blood* 57:1025.

Nakai GS, Craddock CG, Figueroa WG. 1962. Agnogenic myeloid metaplasia: a survey of twenty-nine cases and review of the literature. *Ann Intern Med* 57:419.

Nathan DG, Berlin NI. 1959. Studies on the production and life span of erythrocytes in myeloid metaplasia. *Blood* 14:668.

Nathan DG, Chess L, Hillman DG, Clarke B, Breard J, Merler E, Housman DE. 1978*a*. Human erythroid burst-forming unit: T-cell requirement for proliferation in vitro. *J Exp Med* 147:324.

Nathan DG, Clarke BJ, Hillman DG, Alter BP, Housman DE. 1978*b*. Erythroid precursors in congenital hypoplastic (Diamond-Blackfan) anemia. *J Clin Invest* 61:489.

Nathan DG, Hillman DG. 1978. Studies of erythropoiesis in culture. *Blood Cells* 4:219.

Nathan DG, Hillman DG, Chess L, Alter BP, Clarke BJ, Breard J, Housman DE. 1978*c*. Normal erythropoietic helper T-cells in congenital hypoplastic (Diamond-Blackfan) anemia. *N Engl J Med* 298:1049.

Neame PB, Simpson JC. 1964. Acute transitory erythroblastopenia in kwashiorkor. *S Afr J Lab Clin Med* 10:27.

Necheles TF. 1971. *Proceedings of the Symposium on Androgens in the Anemia of Bone Marrow Failure*. Palo Alto, California: Snytex Lab.

Negri-Gualdi C. 1977. L'anemia cronica a impronta arigenerativa quale possibile forma transizione tra le aplasie midollari globali primitive e le eritroblastopenie idiopatiche croniche non associate a timoma. *Minerva Med* 68:1057.

Newland AC, Catovsky D, Linch D, Cawley JC, Beverly P, San Miguel JF, Gordon Smith EC, Blecher TE, Shahriari S, Varadi S. 1984. Chronic T cell lymphocytosis: a review of 21 cases. *Br J Haematol* 58:433.

Nicholls MD, Concannon AJ. 1982. Maloprim-induced agranulocytosis and red cell aplasia. *Med J Aust* 2:564.

Nicrosini F, Taraschi G. 1965. Su di un caso di tumore del timo complicatosi di polimiosite e di aplasia eritroblastica. *Sist Nerv* 17:22.

Nienhuis AW. 1981. Vitamin C and iron. *N Engl J Med* 304:170.

Nixon AD, Buchanan JG. 1967. Hemolytic anemia due to pyruvate kinase deficiency. *NZ Med J* 66:859.

Norman AG. 1965. Thymic tumors with red cell aplasia. *Thorax* 20:193.

Nussenblatt RB, Palestine AG, Rook AH, Scher I, Walker WC, Gery I. 1983. Treatment of intraocular inflammatory disease with cyclosporin-A. *Lancet* ii:235.

O'Brien RT. 1974. Ascorbic acid enhancement of desferrioxamine-induced urinary iron excretion in thalassemia major. *Ann NY Acad Sci* 232:221.

Oeltgen EE, Pribilla W. 1964. Die Erythrokinetik bei Osteomyelofibrose. *Klin Wochenschr* 42:483.

Oie BK, Hertel J, Seip M, Friis-Hansen, B. 1984. Hydrops foetalis in three infants of a mother with acquired chronic pure red cell aplasia: transitory red cell aplasia in one of the infants. *Scand J Haematol* 33:466.

Old CW, Flannery EP, Grogan TM, Stone WH, San Antonio RP. 1978. Azathioprine-induced pure red cell aplasia. *JAMA* 240:552.

Oren ME, Cohen MS. 1978. Immune thrombocytopenia, red cell aplasia, lupus and hyperthyroidism. *South Med J* 71:1577.

Ortega JA, Shore NA, Dukes PP, Hammond D. 1975. Congenital hypoplastic anemia: inhibition of erythropoiesis by sera from patients with congenital hypoplastic anemia. *Blood* 45:83.

Owren PA. 1948. Congenital hemolytic jaundice, pathogenesis of the hemolytic crisis. *Blood* 3:231.

Ozawa K, Kurtzman G, Young N. 1986. Replication of the B19 parvovirus in human bone marrow cell cultures. *Science* 233:883.

Ozer FL, Truax WE, Levin WC. 1960. Erythroid hypoplasia associated with chloramphenicol therapy. *Blood* 16:997.

Özsoylu S. 1984. High-dose intravenous corticosteroid for a patient with Diamond-Blackfan syndrome refractory to classical prednisone treatment. *Acta Haematol* 71:207.

Papayannopoulou T, Vichnisky E, Stamatoyannopoulos G. 1980. Fetal Hb production during acute erythroid expansion: I. observations in patients with transient erythroblastopenia and post phlebotomy. *Br J Haematol* 44:535.

Parry EHO, Kilpatrick GS, Hardisty RM. 1959. Red cell aplasia and benign thymoma: studies on a case responding to prednisone. *Br Med J* i:1154.

Pattison JR, Jones SE, Hodgson J, Davis LR, White JM, Stroud CE, Murtaza L. 1981. Parvovirus infections and hypoplastic crises in sickle cell anemia. *Lancet* 1:664.

Paver WK, Clarke SK. 1976. Comparison of human fecal and serum parvo-like viruses. *J Clin Microbiol* 4:67.

Pavlic GJ, Bouroncle BA. 1965. Megalobastic crisis in paroxysmal nocturnal hemoglobinuria. *N Eng J Med* 273:789.

Penn CR, Hope-Stone HF. 1972. The role of radiotherapy in management of malignant thymoma. *Br J Surg* 59:533.

Perlmann P, Perlmann H, Wigsell H. 1972. Lymphocyte mediated cytotoxicity in vitro. Induction and inhibition by humoral antibody and nature of effector cells. *Transplant Rev* 13:91.

Peschle C. 1980. Erythropoiesis. *Ann Rev Med* 31:303.

Peschle C, Condorelli M. 1972. Eritroblastopenia pura associata a timoma: Dimonstrazione di un agente sierico autoimmuno inibente l'eritropoiesi e remissione indotta. *Hematologica* 58:553.

Peschle C, Marmont AM, Marone G. 1975*a*. The IgG serum inhibitor in adult pure red cell aplasia: assay techniques and mechanism of action. In *Erythropoiesis,* ed. K Nanao, JW Fisher, F. Takaku. Baltimore: University Park Press, 489.

Peschle C, Marmont AM, Marone G, Genovese A, Sasso G, Condorelli M. 1975*b*. Pure red cell aplasia: studies on an IgG serum inhibitor neutralizing erythropoietin. *Br J Haematol* 30:411.

Peschle C, Marmont A, Perugini S, Bernasconi C, Brunetti P, Fontana G, Ghio R, Resegotti L, Rizzo SC, Condorelli M. 1978. Physiopathology and therapy of adult pure red cell aplasia: a cooperative study. In *Aplastic Anemia*, ed. S Hibino. Baltimore: University Park Press, 285.

Pezzimenti JF, Lindenbaum J. 1972. Megaloblastic anemia associated with erythroid hypoplasia. *Am J Med* 53:748.

Pierre RV. 1974. Preleukemic states. *Semin Hematol* 11:73.

———. 1975. Cytogenetic studies in preleukemia: studies before and after transition to acute leukemia in 17 subjects. *Blood Cells* 1:163.

Pirofsky B. 1969. Autoimmunity and the Autoimmune Hemolytic Anemias. Baltimore: Williams & Wilkins, 86.

———. 1976. Clinical aspects of autoimmune hemolytic anemia. *Semin Hematol* 13:251.

Pisciotta AV, Hinz JE. 1956. Occurrence of agglutinins in normoblasts. *Proc Soc Exp Biol Med* 91:356.

Planas AT, Kranwinkel RN, Soletsky HG, Pezzimenti JF. 1980. Chlorpropamide-induced pure RBC aplasia. *Arch Intern Med* 140:707.

Polayes SH, Lederer M. 1930. Recipient of many blood transfusions. *JAMA* 95:407.

Pollack S, Cunningham-Rundles C, Smithwick EM, Barandum S, Good RA. 1982. High-dose intravenous gamma globulin for autoimmune neutropenia. *N Engl J Med* 307:253.

Pollycove M, Mortimer R. 1961. The quantitative determination of iron kinetics and hemoglobin synthesis in human subjects. *J Clin Invest* 40:753.

Pollycove M, Tono M. 1975. Studies of the erythron. *Semin Nucl Med* 5:11.

Potter CG, Potter AC, Hatton CSR, Chapel HM, Anderson MJ, Pattison JR, Tyrrell DA, Higgins PG, Willman JS, Parry HF, Cotes PM. 1987. Variation of erythroid and myeloid precursors in the marrow and peripheral blood of volunteer subjects with human parvovirus (B19). *J Clin Invest* 79:1486.

Prasad AS, Berman L, Tranchida L, Poulik MD. 1968. Red cell hypoplasia, cold hemoglobinuria and M-type gamma G serum paraprotein and Bence-Jones proteinuria in a patient with lymphoproliferative disorder. *Blood* 31:151.

Prasad AS, Tranchida L, Palutke M, Ponlik DM. 1976. Red cell hypoplasia and monoclonal gammopathy in a patient with lymphoproliferative disorder. *Am J Med Sci* 271:355.

Price JM, Brown RR, Pfaffenbach EC, Smith NJ. 1970. Excretion of urinary tryptophan metabolites in patients with congenital hypoplastic anemia. *J Lab Clin Med* 75:316.

Pritchard KI, Quirt IC, Simpson WJ, Fleming JF. 1979. Phenytoin-associated reversible red cell aplasia. *Can Med Assoc J* 121:1491.

Propper RD, Cooper B, Rufo RR, Nienhuis AW, Anderson WF, Bunn HF, Rosenthal A, Nathan DG. 1977. Continuous subcutaneous administration of deferoxamine in patients with iron overload. *N Engl J Med* 297:418.

Propper RD, Shurin SB, Nathan DG. 1976. Reassessment of the use of desferrioxamine-B in iron overload. *N Engl J Med* 294:1421.

Purtilo DT, Zelkowitz L, Harada S, Brooks CD, Bechtold T, Lipscomb H, Yetz J, Rogers G. 1984. Delayed onset of infectious mononucleosis associated with acquired agammaglobulinemia and red cell aplasia. *Ann Intern Med* 101:180.

Radermecker M, Oger A, Lambert PH, Messens Y. 1964. Erythroblastopénie et tumeur du thymus: correction remarquable de l'anémie par la corticotherapie. *Presse Med* 72:1115.

Ramos AJ, Loeb V. 1956. Diagnostic problems: presentation of a case and comments. *JAMA* 160:1317.

Rao KRP, Patel AR, Anderson MJ, Hodgson J, Jones SE, Pattison DM. 1983. Infection with parvovirus-like virus and aplastic crisis in chronic hemolytic anemia. *Ann Intern Med* 98:930.

al-Rashid RA. 1979. Idiopathic transient erythroblastopenia of childhood. *Am J Pediatr Hematol Oncol* 1:363.

Recker RR, Hynes HE. 1969. Pure red blood cell aplasia associated with chlorpropamide therapy. Patient summary and review of the literature. *Arch Intern Med* 123:445.

Reid G, Patterson AC. 1977. Pure red-cell aplasia after gold treatment. *Br Med J* 2:1457.

Reitz CL, Bottomley SS. 1984. Pure red cell aplasia associated with fenoprofen. *Am J Med Sci* 287:62.

Resegotti L, Dolci C, Palestro G, Peschle C. 1978. Paraproteinemic variety of pure red cell aplasia: immunological studies in 1 patient. *Acta Haematol* 60:227.

Resegotti L, Ricci C. 1976. IgG paraproteinemia in a patient with pure red cell aplasia. In *Leukemia and Aplastic Anemia*, ed. D Metcalf, M Condorelli, C Peschle. Rome: Il Pensiero Scientifico, 48.

Ricci JA, Scott JH, King MH, Glynn MFX, Freedman MH. 1980. Acquired inhibitor of erythropoiesis: red cell failure with marrow erythroid hyperplasia. *Proceedings of the 18th Congress of the International Society of Hematology*. Montreal, (abstract no. 404), 95.

Richet G, Alagille D, Fournier E. 1954. L'érythroblastopénie aiguë de l'anurie. *Presse Med* 62:50.

Ringertz N, Lidholm SO. 1956. Mediastinal tumors and cysts. *J Thorac Surg* 31:458.

Ritchey AK, Hoffman R, Dainiak N, McIntosh S, Weininger R, Pearson HA. 1979. Antibody mediated acquired sideroblastic anemia: response to cytotoxic therapy. *Blood* 54:734.

Ritter J, Zeller H. 1977. Die akute benigne Erythroblastopenie mit normochromer Anämie, eine nicht seltene Erkrankung im Kleinekidersalter. *Monatsschr Kinderheilkd* 124:44.

River GL. 1966. Erythroid aplasia following thymectomy: report of a case with positive lupus erythematosus cell preparation and elevated plasma erythropoietin level. *JAMA* 197:726.

Roberts HJ. 1983. Aplastic anemia and red cell aplasia due to pentachlorophenol. *South Med J* 76:45.

Robins-Browne RM, Green R, Katz J, Becker D. 1977. Thymoma, pure red cell aplasia, pernicious anemia and candidiasis: a defect in immunohomeostasis. *Br J Haematol* 36:5.

Rogers BH, Manligod JR, Blazek WV. 1968. Thymoma associated with pancytopenia and hypogammaglobulinemia: report of a case and review of the literature. *Am J Med* 44:154.

Rohr K. 1949. Zur Pathogenese der Erythroblastischen Markinsuffizienz. *Verh Dtsch Ges Inn Med* 58:666.

Roland AS. 1964. The syndrome of benign thymoma and primary aregenerative anemia; an analysis of 43 cases. *Am J Med Sci* 247:719.

Romano E, Layrisse M, Romano M, Soyano A, Layrisse Z. 1980. Electron microscopic demonstration of IgG antibodies directed to erythroblast in primary acquired pure red cell aplasia. *Clin Immunol Immunopathol* 17:330.

Rosai J. Levine GD. 1976. Tumors of the thymus. In *Atlas of Tumor Pathology*, ed.

HI Firminger. Washington, D.C.: Armed Forces Institute of Pathology, Fasc 13, 31.

Ross JF, Finch SC, Street RE, Jr, Streider JW. 1954. The simultaneous occurrence of benign thymoma and refractory anemia. *Blood* 9:935.

Rossi W, Diena F, Sacchetti C. 1957. Demonstration of specific agglutinogens in normal bone marrow erythroblasts. *Experientia* 13:440.

Rubin M, Straus B, Allen L. 1964. Clinical disorders associated with thymic tumors. *Arch Intern Med* 114:389.

Rubin RN, Walker BK, Ballas SK, Travis SF. 1978. Erythroid aplasia in juvenile rheumatoid arthritis. *Am J Dis Child* 132:760.

Rullan-Ferrer JA, Marchand EJ, de Torregrosa MV. 1955. Relationship of pancytopenia to megaloblastic anemias: report of seven cases clinically confused with blood dyscrasias and subacute bacterial endocarditis. *JAMA* 157:638.

Sacks PV. 1974. Autoimmune hematologic complications in malignant lymphoproliferative disorders. *Arch Intern Med* 134:781.

Safdar SH, Krantz SB, Brown EB. 1970. Successful immunosuppressive treatment of erythroid aplasia appearing after thymectomy. *Br J Haematol* 19:435.

Saint-Aimé C. 1971. Erythroblastopénie chronique idiopathique chez trois nourrissons. *Arch Fr Pediatr* 28:991.

Sakol MJ. 1954. Red cell aplasia. *Arch Intern Med* 94:481.

Salama A, Mueller-Eckhardt C, Kiefel V. 1983. Effect of intravenous immunoglobulin in immune thrombocytopenia: competitive inhibition of reticuloendothelial system function by sequestration of autologous red blood cells. *Lancet* ii:193.

Sallan SE, Buchanan GR. 1977. Selective erythroid aplasia during therapy for acute lymphoblastic leukemia. *Pediatrics* 59:895.

Sanz MA, Martinez JA, Gomis F, Garcia-Borras JJ. 1980. Sulindac-induced bone marrow toxicity. *Lancet* 2:802.

Sao H, Yoshikawa H, Akao Y, Hiraiwa A, Takagi S, Yamauchi T, Yoshikawa S. 1982. An adult case of acute erythroblastopenia. *Rinsho Ketsueki* 23:1453.

Sass MD, Vorsanger E, Spear W. 1964. Enzyme activity as an indicator of red cell age. *Clin Chim Acta* 10:21.

Savage RA. 1976. Pure red cell aplasia: a preleukemic state. *Cleve Clin Q* 43:267.

Sawada K. Koyanagawa Y, Sakurama S, Nagakawa S, Konno T. 1985. Diamond-Blackfan syndrome: a possible role of cellular factors for erythropoietin suppression. *Scand J Haematol* 35:158.

Schafer AI, Rabinows S, Le Boff MS, Bridges K, Cheron RG, Dluhy R. 1985. Long-term efficacy of deferoxamine iron chelation therapy in adults with acquired transfusional iron overload. *Arch Intern Med* 145:1217.

Schmid JR, Kiely JM, Harrison EG, Bayrd ED, Pease GL. 1965. Thymoma associated with pure red cell agenesis: review of literature and report of 4 cases. *Cancer* 18:216.

Schmid JR, Kiely JM, Pease GL, Hargraves MM. 1963. Acquired pure red cell agenesis: report of 16 cases and review of the literature. *Acta Haematol* 30:255.

Schneerson JM, Mortimer PP, Vandervelde EM. 1980. Febrile illness due to a parvovirus. *Br Med J* 280:1580.

Schneider H, Jobke A, Niederhoff H, Kunzer W. 1985. Passagere Erythroblastopenie. *Klin Padiatr* 197:9.

Schooley JC, Garcia JF. 1962. Immunochemical studies of human urinary erythropoietin. *Proc Soc Exp Biol Med* 109:325.

———. 1965. Some properties of serum obtained from rabbits immunized with human urinary erythropoietin. *Blood* 25:204.

Schorr JB, Cohen ES, Schwarz A, Wallerstein H. 1960. Hypoplastic anemia in childhood. *Jewish Mem Hosp Bull* 5:126.

Schreiber AD. 1980. Systemic lupus erythematosus: hematological aspects. *J Rheumatol* 7:395.

Schroeder WA, Huisman THJ, Brown AK. 1971. Postnatal changes in the chemical heterogeneity of human fetal hemoglobin. *Pediatr Res* 9:493.

Schulof RS, Goldstein AL. 1977. Thymosin and the endocrine thymus. *Adv Intern Med* 22:121.

Seaman AJ, Kohler RD. 1953. Acquired erythrocytic hypoplasia: a recovery during cobalt therapy. *Acta Haematol* 9:153.

Sears DA, George JN, Gold MS. 1974. Transient red blood cell aplasia in association with viral hepatitis. *Arch Intern Med* 135:1585.

Seewann HL. 1979. Subakute idiopatische autoimmunhaemolytische Anämie mit protrachierter aplatisher Phase und erythramischer Reaktion. *Wien Med Wochenschr* 129:180.

Seidenfeld AM, Owen J, Prchal JF, Glynn MF. 1979. Pure red cell aplasia with an inhibitor to erythropoiesis. *Can Med Assoc J* 121:188.

Seip M. 1955. Aplastic crisis in a case of immunohemolytic anemia. *Acta Med Scand* 153:137.

Seligmann M, Bernard J, Chassigneux J, Dresch C. 1963. Anémie hypoplastique familiale. *Nouv Rev Fr Haematol* 3:209.

Serjeant GR, Mason K, Topley JM, Serjeant BE, Pattison JR, Jones SE, Mohamed R. 1981. Outbreak of aplastic crises in sickle cell anemia associated with parvovirus-like agent. *Lancet* 2:595.

Sharff O, Neumann H. 1944. Über eine seltene Form von Knochenmarkschadidung durch Salvarsan. *Med Klin* 40:500.

Shevach EM. 1985. The effects of cyclosporin A on the immune system. *Ann Rev Immunol* 3:397.

Shillitoe AJ, Goodyear JE. 1960. Thymolipoma: a benign tumor of the thymus gland. *J Clin Pathol* 13:297.

Shimm DS, Logue GL, Rohlfing MB, Gaede JT. 1979. Primary amyloidosis, pure red cell aplasia and Kaposi's sarcoma in a single patient. *Cancer* 44:1501.

Shionoya S, Amano M, Imamura Y, Nakahara K, Okawa H. 1984. Suppressor T-cell chronic lymphocytic leukemia associated with red cell hypoplasia. *Scand J Haematol* 33:231.

Shiraishi Y, Yamamoto K, Taguchi H, Ueda N, Shiomi F. 1980. Philadelphia chromosome in pure red cell aplasia: a preleukemic state? *Cancer Genet Cytogenet* 2:1.

Signier F, Betourne G, Mathé G, Caquet R, Milhaud A, Kahn F. 1960. L'association de tumeur thymique et aplasie médullaire. *Sem Hop Paris* 36:165.

Singer K, Motulsky AG, Wile SA. 1950. Aplastic crisis in sickle-cell anemia: study of its mechanism and its relationship to other types of hemolytic crisis. *J Lab Clin Med* 35:721.

Sjölin S, Wranne L. 1970. Treatment of congenital hypoplastic anemia with prednisone. *Scand J Haematol* 7:63.

Skikne BS, Lynch SR, Bezwoda WR. 1976. Pure red cell aplasia. *S Afr Med J* 50:1353.

Slater LM, Schlutz MJ, Armentrout SA. 1979. Remission of pure red cell aplasia associated with non thymic malignancy. *Cancer* 44:1879.

Smith CH. 1949. Chronic congenital aregenerative anemia (pure red cell anemia) associated with iso-immunization by blood group factor "A". *Blood* 4:697.

Socinski MA, Ershler WB, Frankel JP, Albertini RJ, Ciongoli AK, Krawitt EL, Burns SL, Mangan KF. 1983. Pure RBC aplasia and myasthenia gravis: coexistence of two diseases associated with thymoma. *Arch Intern Med* 143:543.

Socinski MA, Ershler WB, Tosato G, Blaese RM. 1984. Pure red blood cell aplasia associated with chronic Epstein-Barr virus infection: evidence for T-cell-mediated suppression of erythroid colony-forming units. *J Lab Clin Med* 104:995.

Soler J, Estivill X, Ayats R, Brunet S, Pujol-Moix N. 1985. Chronic T-cell lymphocytosis associated with pure red cell aplasia, thymoma and hypogammamglobulinemia. *Br J Haematol* 61:582.

Souadjian JV, Enrique P, Silverstein MN, Pepin JM. 1974. The spectrum of diseases associated with thymoma: coincidence or syndrome? *Arch Intern Med* 134:374.

Soulter L, Emerson CP. 1960. Elective thymectomy in the treatment of aregenerative anemia associated with monocytic leukemia. *Am J Med* 28:609.

Soulter L, Sommers S, Relman AS, Emerson CP. 1957. Problems in the surgical management of thymic tumors. *Ann Surg* 146:424.

Souquet R, Batime J, Kermarec J, Hagnenauer G, Teyssier L, Pernod J. 1970. Erythroblastopénie et déficit en immunoglobulines après cobaltotherapie d'un thymome évoluant depuis 26 ans. *Poumon et Coeur* 26:1203.

Spiers ASD. 1976. The treatment of chronic granulocytic leukaemia. *Br J Haematol* 32:291.

Starling KA, Fernbach DJ. 1967. Congenital hypoplastic anemia: a review of eight cases. *Tex Med* 63:63.

Steffen C. 1955. Untersuchungen über den Nachweiss sessiler Antikorper an Knockenmarkzellen bei erworbener hämolytischer Anämie. *Wien Klin Wochenschr* 67:224.

Steinberg MH, Coleman MF, Pemebaker JB. 1979. Diamond-Blackfan syndrome: evidence for T-cell-mediated suppression of erythroid development and a serum-blocking factor associated with remission. *Br J Haematol* 41:57.

Stephens ME. 1974. Transient erythroid hypoplasia in a patient on long-term co-trimoxazole therapy. *Postgrad Med J* 50:235.

Stiller CR, Dupre J, Gent M, Jenner MR, Keown PA. 1984. Effect of cyclosporin immunosuppression in insulin-dependent diabetes mellitus of recent onset. *Science* 223:1362.

Stohlman F, Quesenberry PJ, Howard D, Miller ME, Schur P. 1971. Erythroid aplasia, an autoimmune complication of chronic lymphocytic leukemia. *Clin Res* 19:566.

Strauss AM. 1943. Erythrocyte aplasia following sulfathiazole. *Am J Clin Pathol* 13:249.

Sugimoto M, Wakabayashi Y, Shiokawa Y, Takaku F. 1982. A case of pure red cell aplasia whose bone marrow heme synthesis was suppressed by diphenylhydantoin in vitro. *Rinsho Ketsueki* 23:1446.

Sultan Y, Maisonneuve P, Kazatchkine MD, Nydegger UE. 1984. Anti-idiotypic suppression of autoantibodies to factor VIII (antihaemophilic factor) by high-dose intravenous gammaglobulin. *Lancet* ii:765.

Swineford O, Curry JC, Cumbia JW. 1958. Phenylbutazone toxicity: depression of erythropoiesis: a case report. *Arthritis Rheum* 1:174.

Szur L, Smith MD. 1961. Red cell production and destruction in myelosclerosis. *Br J Haematol* 7:147.

Takigawa M, Hayakawa M. 1974. Thymoma with systemic lupus erythematosus, red blood cell aplasia and herpes virus infection. *Arch Dermatol* 110:99.

Talerman A, Amigo A. 1968. Thymoma associated with aregenerative and aplastic anemia in a five-year-old child. *Cancer* 21:1212.

Tartaglia AP, Propp S, Amarose AP, Propp PR, Hall AC. 1966. Chromosome abnormality and hypocalcemia in congenital erythroid hypoplasia (Diamond-Blackfan syndrome). *Am J Med* 41:990.

Tatarsky I. 1972. Transient erythroid hypoplasia in chronic lymphatic leukemia. *N Instabul Contrib Clin Sci* 10:130.

Tattersall P, Ward D., eds. 1978. *The Parvoviruses: An Introduction in Replication of Mammalian Parvoviruses.* Cold Spring Harbor, New York: Cold Spring Harbor Laboratory, 3.

Teoh PC, Tan DKS, Da Costa JL, Chew BK. 1973. Acquired pure red cell aplasia in adults. *Med J Aust* 2:373.

Thevénard A, Marquès JM. 1955. Myasthénie, tumeur de thymus et aplasie (ou hypoplasie) de la moelle osseuse: a propos d'un cas suivis depuis treize ans. *Rev Neurol* 93:597.

Thiagarajan D, Tawadros H, Dipaling S. 1983. Pure red blood cell aplasia complicating chronic lymphatic leukemia. *Am J Med Sci* 286:22.

Tiber C, Casimir M, Nogeire C, Lichtiger B, Conrad FG. 1981. Thymoma with red cell aplasia and hemolytic anemia. *South Med J* 74:1164.

Tillmann W, Prindull G, Schroter W. 1976. Severe anemia due to transient pure red cell aplasia in early childhood: arrest at the level of the committed stem cells. *Eur J Pediatr* 123:51.

Toogood IR, Speed IE, Cheney KC, Rice MS. 1978. Idiopathic transient normocytic normochronic anemia of childhood. *Aust Paediatr J* 14:28.

Toole JF, Witcofski R. 1966. Selenomethionine-Se^{75} scan for thymoma. *JAMA* 198:1219.

Torok-Storb B, Hansen JA. 1982. Modulation of in vitro BFU-E growth by normal Ia-positive T cells is restricted by HLA-DR. *Nature* 298:473.

Torok-Storb B, Martin PJ, Hansen JA. 1981. Regulation of in vitro erythropoiesis by normal T cells: evidence for two T-cell subsets with opposing function. *Blood* 58:171.

Torok-Storb BJ, Siell G, Storb R, Adamson J, Thomas ED. 1980. In vitro tests for distinguishing possible immune-mediated aplastic anemia from transfusion-induced sensitization. *Blood* 55:211.

Tosato G, Magrath I, Koski I, Dooley N, Blaese M. 1979. Activation of suppressor T cells during Epstein-Barr virus-induced infectious mononucleosis. *N Engl J Med* 301:1133.

Tötterman TH, Nissel J, Killander A, Gahrton G, Lönnqvist B. 1984. Successful treatment of pure red cell aplasia with cyclosporin. *Lancet* 2:693.

Trainin N. 1974. Thymic hormones and the immune response. *Physiol Rev* 54:272.

Tsai SY, Levin WC. 1957. Chronic erythrocytic hypoplasia in adults: review of literature and report of a case. *Am J Med* 22:322.

Tsubanio T, Kurata Y, Katagiri S. 1983. Alteration of T-cell subsets and immunoglobulin synthesis in vitro during high-dose gamma globulin therapy in patients with idiopathic thrombocytopenic purpura. *Clin Exp Immunol* 53:697.

Turpin F, Farhat M. 1981. Erythroblastopénie primitive à rechute de l'adulte: possible rôle suppressive des cellules mononucléées sanguines sur l'érythropoïèse. *Rev Med Interne* 2:395.

Varet B, Casadevall N, Etevenaux J, Chistoforov B, Pequignot H. 1978. Erythroblastopénie apres thymectomie: guérison après traitement par la cyclophosphamide. *Ann Med Interne* 129:509.

Vasavada PJ, Bournigal LJ, Reynolds RW. 1973. Thymoma associated with pure red cell aplasia and hypogammaglobulinemia. *Postgrad Med* 54:93.

Viala JJ, Ville D, Coiffier B, Rebattu P, Guastalla JP. 1981. Les érythroblastopénies de la leucémie lymphoide chronique: intérêt de la cinétique du radiofer. *Nouv Rev Fr Haematol* 23:213.

Vilan J, Rhyner K, Ganzoni A. 1971. Immunosuppressive treatment of pure red cell aplasia. *Lancet* 2:51.

———. 1973. Pure red cell aplasia: successful treatment with cyclophosphamide. *Blut* 26:27.

Vilter RW, Jarrold T, Will JJ, Mueller JF, Friedman BI, Hawkins VR. 1960. Refractory anemia with hyperplastic bone marrow. *Blood* 15:1.

Vodopick H. 1975. Cherchez la chienne: erythropoietic hypoplasia after exposure to gamma-benzene hexachloride. *JAMA* 234:850.

Vohra RM, Sturm RE, Patel AR. 1983. Pure red cell aplasia in association with allopurinol. *Blood* 62(suppl 1):51a.

Voyce MA. 1963. A case of pure red cell aplasia successfully treated with cobalt. *Br J Haematol* 9:412.

Walt F, Taylor JE, Magill FB, Nestadt A. 1962. Erythroid hypoplasia in kwashiorkor. *Br Med J* i:73.

Wang WC,Mentzer WC. 1976. Differentiation of transient erythroblastopenia of childhood from congenital hypoplastic anemia. *J Pediatr* 88:784.

Wara DW, Goldstein AL, Doyle W, Ammann AJ. 1975. Thymosin activity in patients with cellular immunodeficiency. *N Engl J Med* 292:70.

Ward HP, Block MH. 1971. The natural history of agnogenic myeloid metaplasia (AMM) and a critical evaluation of its relationship to the myeloproliferative syndrome. *Medicine* 50:357.

Wasi P, Block MM. 1961. The mechanism of the development of anemia in untreated chronic lymphocytic leukemia. *Blood* 17:597.

Wasser JS, Yolken R, Miller RD, Diamond L. 1978. Congenital hypoplastic anemia (Diamond-Blackfan syndrome) terminating in acute myelogenous leukemia. *Blood* 51:991.

Waterkotte GW, McElfresh AE. 1974. Congenital pure red cell hypoplasia in identical twins. *Pediatrics* 54:646.

Wegelius R, Weber TH. 1978. Transient erythroblastopenia of childhood: a study of 15 cases. *Acta Paediatr Scand* 67:513.

Weinbaum JG, Thomson RF. 1955. Erythroblastic hypoplasia associated with thymic tumor and myasthenia gravis. *Am J Clin Pathol* 25:761.

Weinberger KA. 1979. Fenoprofen and red cell aplasia. *J Rheumatol* 6:475.

Weiner M, Karpatkin M, Hart D, Seaman C, Vora SK, Henry WL, Piomelli S. 1978. Cooley anemia: high transfusion regimen and chelation therapy, results and perspective. *J Pediatr* 92:653.

Wendt F, Doyen A, Schoop W, Schubothe H, Hunstein H, Fliendner TM, Wedler HW, 1965. Erythroblastophthise bei Thymom: Bericht uber zwei beobachtungen mit Remission. *Schweiz Med Wochenschr* 95:1494.

Whitaker JA, Vogler WR, Werner JH. 1970. Red cell aplasia and osteoblastic metastases in a patient with thymoma. *Cancer* 26:427.

Whitby LEH, Britton CJC. 1953. *Disorders of the Blood.* 7th edition. New York: Grune and Stratton.

Wibulyachainunt S, Price P, Brill AB, Krantz SB. 1978. Studies on red cell aplasia: IX. ferrokinetics during remission of the disease. *Am J Hematol* 4:233.

Wilkins EW, Castleman B. 1979. Thymoma: a continuing survey at the Massachusetts General Hospital. *Ann Thorac Surg* 28:252.

Wilson HA, McLaren GD, Dworken HJ, Tebbi K. 1980. Transient pure red cell aplasia: cell-mediated suppression of erythropoiesis associated with hepatitis. *Ann Intern Med* 92:196.

Wranne L. 1970. Transient erythroblastopenia in infancy and childhood. *Scand J Haematol* 7:76.

Wunderlich J, Rosenberg E, Connolly J, Parks J. 1975. Characteristics of a cytotoxic human lymphocyte-dependent antibody. *J Natl Cancer Inst* 54:537.

Yoo D, Pierce LE, Lessin LS. 1983. Acquired pure red cell aplasia associated with chronic lymphocytic leukemia. *Cancer* 51:844.

Young NS, Harrison M, Mortimer PP, Moore JG, Humphries RK. 1984*a*. Direct demonstration of the human parvovirus in erythroid progenitor cells infected in vitro. *J Clin Invest* 74:2024.

Young NS, Mortimer PP. 1984. Viruses and bone marrow failure. *Blood* 63:729.

Young NS, Mortimer PP, Moore JG, Humphries RK. 1984*b*. Characterization of a virus that causes transient aplastic crisis. *J Clin Invest* 73:224.

Yunis AA, Arimura GK, Lutcher CL, Blasquez J, Halloran M. 1967. Biochemical lesion in dilantin-induced erythroid aplasia. *Blood* 30:587.

Yunis AA, Bloomberg GR. 1964. Chloramphenicol toxicity: clinical features and pathogenesis. In *Progress in Hematology*, vol. 4, ed. EB Brown, CV Moore. New York: Grune and Stratton, 138.

Yunis JJ, Yunis EJ. 1963. Cell antigens and cell specialization: I. a study of blood group antigens on normoblasts. *Blood* 22:53.

———. 1964. Cell antigens and cell specialization: II. demonstration of some red cell antigens on human normoblasts. *Blood* 24:522.

Zaentz SD, Krantz SB. 1973. Studies on pure red cell aplasia. VI: development of two-stage erythroblast cytotoxicity method and role of complement. *J Lab Clin Med* 82:31.

Zaentz SD, Krantz SB, Brown EB. 1976. Studies on pure red cell aplasia: VIII. maintenance therapy with immunosuppressive drugs. *Br J Haematol* 32:47.

Zaentz SD, Krantz SB, Sears DA. 1975. Studies on pure red cell aplasia: VII. presence of proerythroblasts and response to splenectomy: a case report. *Blood* 46:261.

Zaentz SD, Luna JA, Baker AS, Krantz SB. 1977. Detection of cytotoxic antibody to erythroblasts. *J Lab Clin Med* 89:851.

Zalusky R, Zanjani ED, Gidari AS, Ross J. 1973. Site of action of a serum inhibitor of erythropoiesis. *J Lab Clin Med* 81:867.

Zeok JV, Todd EP, Dillon M, DeSimone P, Utley JR. 1979. The role of thymectomy in red cell aplasia. *Ann Thorac Surg* 28:257.

Zielke HR, Ozand PT, Luddy RE, Zinkham WH, Schwartz AD, Sevdalian DA. 1979. Elevation of pyrimidine enzyme activities in the RBC of patients with congenital hypoplastic anemia and their parents. *Br J Haematol* 42:381.

Zoumbos NC, Ferris WO, Hsu SM, Goodman S, Griffith P, Sharrow SO, Humphries RK, Nienhuis AW, Young N. 1984. Analysis of lymphocyte subsets in patients with aplastic anemia. *Br J Haematol* 58:95.

Zoupanos G. 1970. Thymomes et syndromes associés. *Helv Chir Acta* 37:52.

Zucker JM, Tchernia G, Vuylsteke P, Bechart-Michel R, Giorgi R, Blot I. 1971. Erythroblastopénie aiguë secondaire et transitoire au cours du kwashiorkor traité. *Nouv Rev Fr Haematol* 11:131.

Zucker S, Likhite VV, Weintraub LR, Crosby WH. 1974. Remission in pure red cell aplasia following immunosuppressive therapy. *Arch Intern Med* 134:317.

Zuckerman KS. 1981. Human erythroid burst-forming units: growth in vitro is dependent on monocytes, but not T-lymphocytes. *J Clin Invest* 67:702.

Index